TestGoal

Derk-Jan de Grood

TestGoal

Result-Driven Testing

 Springer

Derk-Jan de Grood
Collis B.V.
De Heijderweg 1
2314 XZ Leiden
The Netherlands
grood@collis.nl

ISBN 978-3-540-78828-7   e-ISBN 978-3-540-78829-4

DOI 10.1007/978-3-540-78829-4

ACM Computing Classification (1998): K.6, D.2.5

Library of Congress Control Number: 2008924999

Cover Design: KünkelLopka, Heidelberg
Illustrations: Thijs Geritz, The Hague
Back cover photo: H. de Vries, Rijswijk

Printed on acid-free paper

9 8 7 6 5 4 3 2 1

springer.com

Preface by Lee Copeland

Focus on business goals. Align your work with those goals. Eliminate work that does not add value to the business – this is today's management mantra. All good advice. But few in the testing community truly understand what that kind of alignment means. Derk-Jan de Grood is one of those few.

Today, many variations of testing processes are available to organizations. Some are tool driven (both commercial and open source); others are document driven (IEEE 829 Standard for Software Test Documentation); while still others are technique driven (boundary value, state-transition, and pair-wise testing). A myriad of books are available to help you from the classics by Beizer and Myers to the latest from Black; Bach, Kaner, and Pettichord; Graham, Evans, and van Veenendaal; Craig; and Copeland.

TestGoal is different. TestGoal is **result-driven**. Not the kind of results testers have historically tried to achieve–find all the severity 1 defects, reach 100% statement coverage, or accurately estimate the number of defects remaining. TestGoal focuses on results that the **business** cares about. Like it or not, the business does not care about pair-wise test design and defect taxonomies and defect reports no matter how pretty our charts and graphs may be. The business cares about business results – sales, profit, market share, time to market, product differentiation, and competitive advantage.

As testers, our goals must not only support the goals of the business, they must *be* the goals of the business. Result-driven testing understands those goals, carries out *only* those activities that contribute to those goals, and produces information that enables executive management to achieve those goals. In this way, all of our testing activities support and add value to the organization.

One way TestGoal accomplishes this is by asking us to consider the question "How would you know that this test project has been a success?" How many executive managers would respond, "Well, when you have achieved 90% condition coverage, of course." They don't even know what those words mean. Yet that is what we measure and report – a sign of our disconnectedness from the business. TestGoal puts us on the right track.

Another way TestGoal guides us is by pointing out that the "risks" in risk analysis are those things that prevent the business goals from being achieved – profit, image, product differentiation. Yet most testing risk analysis focuses on threats at a very different level–resource availability, staff availability, and training needs.

TestGoal also emphasizes the central role that testers should play in their organizations. Gone are the days when the testing group was tucked away in the corner of the organization chart. Today's testers play a central role in the creation of quality systems, building bridges to all parts of the organization. We add value to the organization when we influence others in aligning their work to the business.

This book, however, is more than just high-level philosophy. It describes the intricacies of test plans, levels, budgets, strategies, techniques, and designs. It explains the importance of test environments, execution, reporting, and automation. But throughout, Derk-Jan reminds us that while "Testers see a lot of suboptimal processes and faulty systems … This doesn't change the fact that testing is a wonderful and fun profession. Result-driven testers … know they add value to the software development project." It is a great profession. I'm proud to be a part of it. Best wishes in your testing efforts.

March 2008   *Lee Copeland*

Lee Copeland has over thirty years experience as an information systems professional and has held a number of technical and managerial positions with commercial and non-profit organizations in the areas of applications development, software testing, and software development process improvement. He is a well-known and highly regarded speaker at software conferences both in the United States and internationally. Lee is the author of *A Practitioner's Guide to Software Test Design*, a compendium of the most effective methods of test case design.

Preface by Martin Pol

For many organizations, software has become one of the most important pillars of their business processes. The requirements for the timely availability and quality of information systems are high by definition. But in spite of all of the innovations and initiatives to improve processes, it's still not easy to develop error-free software. Testing is, and will continue to be, indispensable to identify poor quality in a timely manner, or in other words, to identify the risks of going live. Learning by error has taught organizations that testing is not only a must but that it contributes positively to the company's business goals.

Structured testing began in the United States in the 1960s, and has since developed into a mature and independent field. Structured testing was first carried out in military environments and in the computer hardware industry. Pioneers such as Beizer, Perry, Parnas, Myers, Hetzel and Gelperin have provided a solid methodical and organizational foundation. Throughout the years, the materials they developed have been included in various standards such as IEEE, ISEB and ISTQB. Over time, these standards have been supplemented with extensions, variations and special applications.

As the result of a study that was conducted by the General Accounting Office and due to political pressure, the Dutch Tax and Customs Services was the first (unbelievable but true) to introduce a comprehensive structured test approach in 1985. As an employee of the Dutch Tax and Customs Services, I was able to build a bridge between the American developments and the Dutch approach. In 1988, this approach was published as the "Testing Manual," which was quickly sold out. In the years following its publication, every self-respecting organization, administrative or other, gladly applied this approach, which is the source of every Dutch testing standard. The Netherlands also have an international reputation to maintain, because their contribution to the worldwide testing community is significant.

Now that testing has conquered a position in information processing, the focus has shifted to its added value and the relationship between costs and profits, and, therefore, to the contribution testing makes to the organization's goals. What damage is prevented by testing? What are the risks? And what are the rewards? In short, the main goal of testing is to minimize risk!

Can testing have more than one goal? It sure can! Testing, of course, is still an activity aimed at finding errors. But with the improvement of methods, techniques, and especially of testing tools, the development of reusable testware, expertise and infrastructure have become valuable by-products. As a result, testing is faster, cheaper and better, which of course has a positive effect on the costs and the time-to-market. TestGoal takes a result-driven approach to testing.

TestGoal is a very valuable addition to today's broad range of available test literature. The "result-driven" approach is hotter than ever. This book's clear explanations pleasantly point the different disciplines involved in the testing process in the right direction: "What is needed to achieve the anticipated goal?"

Experience shows that it is not easy to apply a methodology. Pressures such as deadlines, low budgets or low quality stop many good intentions dead in their tracks. Six clearly described steps lead the reader to the goal – the test result.

Throughout the book, TestGoal provides readers with pointers that enable them to focus on the anticipated goal, the test result, and in fact to minimize risk and produce reusable test products. "Who does what when? How are risks analyzed and processes set up?" Everything has a purpose.

For the TestGoal tester, this result-driven mindset (the ten test principles) is almost innate. A unique element of this book is that every letter focuses on these principles, which I interpret as follows: Professional, Result-driven, Confidence-building, Predictable, Supportive, Flexible and Transparent. The flaws of comparable methodologies are discussed and useful solutions introduced, always with the same goal in mind – the result!

Derk-Jan de Grood has made an exceptional contribution to the world of testing. His drive for result is now available to everyone.

I wish you a lot of fun applying the approach and am confident that this book will enable you to achieve your test goals.

Amersfoort, The Netherlands,
February 2007   *Martin Pol*

Martin Pol is the founder of the first Dutch test approach. He was chairman of the EuroSTAR conference many times, and was chairman of the Dutch association for testers, TestNet, for five years. In 1998, he was the first person to receive the European Testing Excellence Award for his exceptional contribution to the field of testing in Europe.

Acknowledgements

A book does not write itself. A lot of people helped create the TestGoal philosophy described in this book. A word of thanks is therefore more than appropriate; this book would not be what it is without their contribution.

A lot of colleagues contributed to the first version of TestGoal by writing down the experiences they gained with specific activities in a test project. Thanks to their efforts, TestGoal is not only my story, but *our* story. For their efforts, I would like to thank the following people:

Albert Los, Albert Witteveen, Colin Lek, Edward van der Pijl, Frodo Wesseling, Gert de Mooij, Hans van Bommel, Jaap Wijnands, Jos Oudsen, Juul Roelofs, Leonard Hugenholtz, Marcel van Donge, Marco van der Heide, Mario de Boer, Mehdi Jalilivandt, Paul van Leeuwen, Roy Miller Sander Panhuyzen, Wilco Schumacher, Mathilda Banfield and Janneke Klop.

In addition, I would like to thank the following people individually: *Trainee class 2005*: You very actively used the primal version of this book. Your questions provided me with valuable information, which I have used to expand the TestGoal story and make it clearer. *Ronald Lagendijk*: Thank you very much for your review and your special contribution to the chapter on test automation and the section on testing conformity and interoperability. *Martin Storm*: You went through the manuscript at a number of vital moments, and I took great pleasure in our discussions. Your test expertise was an indispensable contribution. I would also like to thank you for creating the layout for the chapter on test automation. *Harry Vroom*: It took me weeks to process your review comments, all of which were justified. Your critical examination was a great help. *George Leih*: I should have known that you would ask a lot of questions instead of commenting! I had to think about some of the questions for a while, but you will find the answers in the book. *Juul Roelofs*: Your expertise in test automation and test data has been a great help. Thank you for thinking along with me and for your review. *Susan Zarakoviti:* When you started your review, you knew nothing about TestGoal, which makes you an ideal reviewer. And even more so because I respect your knowledge of the testing profession. *Jaap Wijnands:* Thank you for your meticulous thoughts during the review. *Jan Rodenburg:* Thank you for your contribution on performance testing. *Jasper Overgaauw:* I appreciate how you thought along and helped put exploratory testing into practice. Thanks go to *Cynthia Maasbommel, Jos Oudsen* and *Herman Rus:* for their indispensable contribution on security testing. Thank you. *Vien Sawer* of *L&L Bunnick*: Thanks to your edit work the text just as I wanted it, understandable and easy to read. I Couldn't have done without you. *Thijs Geritz:* You know that I really like your drawings. Thank you for the pictograms and the other illustrations you made for us. I am happy to have them in my book.

Henk van Dam and *Dirk van den Heuvel:* Thank you for your contribution. Our discussion about the position and content of TestGoal was very intense, but it resulted in a joint product. I also want to thank you for the support and the freedom that you gave me while I was writing. Dirk Jan, the first time we met I still had to learn everything about this intriguing profession. I am glad you introduced me to the world of testing and I am thankful for what you taught me. Without your lessons, this book would not have come into existence.

Erik Petersen, Donna McLeod, Fariba Marvasti, Julie Gardiner and *Thérèse Schoch.* Thanks for your interest and pre-reading the English edition.

And finally, I would like to thank the following people:

Martin Pol and Lee Copeland who wrote the prefaces: Thank you for your words of praise. I am honored that you were took interest in my book. I am thrilled you liked it.

Hilda and Babette: I am aware of the fact that writing a book is an enormous drain on the family quality time. Thanks for understanding and for your patience while I was sitting behind my laptop again.

Leiden, The Netherlands
April 2008   *Derk-Jan de Grood*

Content

Step 1 – Goal

Step 2 – Approach

Step 3 – Design

Step 6 – Assurance

Introduction

Software testing is gaining in popularity. Today, organizations are investing more in the quality of their software and in ways that enable them to control it better. Testing is a very suitable tool that enables companies to do just that.

As testers and consultants, my colleagues and I have visited many different companies over the years. Many of the companies we visited have very mature testing processes, others do not. Although business management and other stakeholders understand that testing is necessary, they are often not aware of its added value. Testers who overlook the added value of testing have difficulty demonstrating it and incorporating it in their software development process. Two aspects seem to function as pitfalls time and time again: the reason to test and the actual framing of the test project. In other words, why are we going to test, and how are we going to do it? The many books that are available on software testing do not spend enough time on these aspects. This is one of the reasons why many companies still consider testing as an activity that is separate from the actual development process. Consequently, the collaboration between testers, business managers, users, developers and other stakeholders is not optimal.

My colleagues and I know that things can be done differently. We know that testing adds value, saves money, minimizes risk and increases overall profitability. There's no denying that testing has a lot of added value.

The added value of testing:

- Testing improves the quality of software products
- Testing accelerates the time-to-market of software products

Testing improves the quality of software products by finding errors and enabling them to be solved before the product is released.

Testing also accelerates the time-to-market of software products by providing insight into the actual operation and performance of the software and reducing uncertainties that business stakeholders have.

Testing supports the development process by building bridges to the other parties involved. Testers work with developers to find a solution that meets the business needs and wishes. Testing increases the confidence stakeholders have in the solution, which creates support and speeds up the decision to go live. The test activities provide insight into the quality of the system and the progress of the development. This information helps steer development activities and thus contributes to an efficient development process.

My intention with this book is to highlight the substantial contribution that testing makes to an organization's business goals. This does, however, require an effort on behalf of both the tester and the organization. It also requires a testing method that focuses on the business goals: result-driven testing.

Result-driven testing is more than just applying a methodology. The success of a project is not determined by the methodology, but by the way the methodology is applied. Result-driven testing, therefore, encompasses several aspects:

- A result-driven mindset
- Testing knowledge
- A result-driven approach that supports the application of result-driven testing

TestGoal is a philosophy that supports all three aspects of result-driven testing. The combination of these three aspects into one philosophy is what makes Test-Goal a fully result-driven approach that enables testers to create transparent and well-structured test projects. This not only benefits the company, it also motivates testers by enabling them to give the company visible added value, namely improved product quality and shorter time-to-market.

This book explains how you can make testing result driven. It explains why testing is important and describes all of the activities involved in testing. This makes it a "GO kit" that enables testers to immediately get started. I hope that this book will make you as enthusiastic as I am about the testing profession and the added value testing has for your organisation and its development process. This being the case, I will have achieved my goal with this book.

TestGoal: A Different View on Testing

TestGoal is not my way of introducing yet another methodology. A number of good methodologies have already been developed and written about. But, like any other profession, testing encompasses more than the simple application of a methodology. After all, strict adherence to a specific methodology is no guarantee for success. Success stems from the mindset, enthusiasm, knowledge and skill of the tester. These factors determine whether a methodology is applied successfully and whether testing takes on a result-driven character.

Collis is a professional testing organization that has been testing software under different conditions at both large and small companies for more than ten years. TestGoal is the result of this extensive experience. TestGoal was developed in the field and contains practical descriptions combined with familiar examples. TestGoal helps you combine the mindset, enthusiasm, knowledge and skills needed to successfully streamline your test project.

- TestGoal provides a concrete proposal for an efficient test approach. It defines the activities that must be included in the test project, the sequence in which they are best run, and the products they produce. The process is efficient because this information enables testers to start testing early on in the project.
- TestGoal enables expectations to be aligned and the customer to formulate its wishes. It stimulates discussion between the tester and the stakeholders about the approach. Involving stakeholders in the determination of the approach builds trust and commitment. It also helps align the stakeholders' different expectations.
- For each of the test activities, the drive to achieve goals and the test principles are integrated in the practical descriptions. Testers not only learn what their tasks are, but also how each task contributes to the anticipated goal. The result-driven tester involves the stakeholders in the test process and ensures that they can see how the quality of the software improves and how the risks of going live are reduced one by one. This builds trust and a basis for the go-live decision.
- TestGoal provides companies with a generic test approach. Together with a clear test report, a generic approach makes the test projects more controllable and comparable to other test projects. This increases the efficiency of the development process and accelerates the time-to-market.

In short: TestGoal is not a methodology, but a philosophy. It's a practical philosophy that gives testers the tools and best practices to make choices and create good test projects. TestGoal is an integrated combination of these two aspects and helps testers think and act in a result-driven way.

The Reason for Testing: What are we Going to Test and Why?

A lot of companies are very keen on testing and have a good, structured methodology, but do not have a concrete goal. Some companies have difficulty explaining why some tests are run and what their goal is. This is, however, a very important part of successfully delivering quality software. If the goal of the test is not clear, the tester will not be able to indicate whether the test produces the desired results or contributes to the organization's business goals. And the latter is crucial to any company. Testers who know what they're doing and what their goal is, are not only able to make the right choices, they are also able to explain their activities and their goals to others.

The first part of this book explains why and how we should focus on the company's business goals. It discusses the role business goals play in a test project and how it helps involve stakeholders in the test activities. The ten test principles are discussed after the introduction to result-driven testing. They help the tester apply the theory with the right mindset and are the driving force behind the tester's actions. What characterizes a result-driven tester and what do their actions contribute to?

Most of the literature about test methodologies covers the testing of applications that are in development, in other words, brand new systems. TestGoal is widely applicable, and can be efficiently and successfully used in maintenance or line organizations to test new programs, as well as changes or enhancements to existing applications. The first part of the book takes a look at various levels of testing and how TestGoal is applied in the different environments. It provides insight into the approach of a test project and the activities involved. Separate chapters are dedicated to the testing of new systems, testing in a maintenance environment, and the testing of performance, security, conformity and interoperability.

Setting up the Test Project: How do we Apply the Methodology?

The second important aspect of testing is the set up of the test project and the approach. More than enough books describe how to test, but because the information is usually very generic, many organizations find it difficult to translate theory into practice. The books do not provide enough information to help the readers make the choices required to apply the theory.

TestGoal clearly describes the philosophy of result-driven testing. This practical approach is the basis for the second part of this book. The TestGoal approach bridges the gap between methodology and practice and provides a practical starting point for planning, setting up, running and finishing test activities. This part of the book is based on the best practices my colleagues and I have gathered throughout the years. It enables readers to choose what they need to set up an efficient test project that suits their specific environment. The TestGoal approach is based on a number of principles:

- A number of decisions have already been made in TestGoal based on best practices. Choices are always made at the cost of genericness, but this usually benefits the concreteness and applicability of the approach.
- Where decisions have not been made, TestGoal suggests a "standard" option.
  Arguments for other options are also provided. This gives readers a starting
  point and support should they choose another option.
- Each activity is described clearly and in detail. If the activity produces a product, the information the product should contain is indicated. If possible, guidelines and examples are provided. For example, the description of the physical
  test design contains tips for the clear formulation of the "goal of the test."
- Each activity is explained using an action plan. The action plan enables readers
  to go through the activities in the test project in a practical sequence, whereby
  some activities can be carried out simultaneously. The action plan clarifies the
  relationship between the activities and prevents activities being forgotten.
- Integrated test principles . Several places in the book refer to relevant test principles. This provides readers insight into the practical application of the principles and into the coherence between the different activities in the test process.
- Situations to which best practices apply are indicated. Tips from the field can
  be adopted by readers or can serve as an inspiration for devising one's own solution.

The set up of the test process is closely linked to the tester's drive to achieve goals
and the test principles. The description of the test approach regularly refers to the
test principles introduced in the first part of the book.

Who is this Book Intended for?

Business Managers and Others Who Have an Interest in Producing High-Quality Software

This book addresses a wide target group. First, it addresses business managers,
project managers, functional managers and other stakeholders who are directly or
indirectly involved in the test process. This book explains what the added value of
testing is and what has to be done to achieve it. Result-driven testing tests and
produces reports on the quality of the product *and* the anticipated goal. It gives
business managers who support result-driven testing immediate insight into the
degree to which the anticipated goal has been achieved and insight into the associated business risks. No lengthy reports, but one or two pages that contain all of the
information business managers need to made well-founded decisions.

Junior and Senior Testers

Second, it addresses the testers. This book gives them insight into the role the goal plays in their activities. Testers must be aware of the added value of tests so they can run them effectively. This book describes a tester's goals as well as the activities he needs to carry out in order to achieve them. It also contains numerous examples and practical tips that junior testers can use as starting points to work out the details of the activities they are asked to perform. Senior testers will also benefit from the book in more way than one. On the one hand, the book helps them streamline their activities, which enables them to work on the organization's view on testing. On the other hand, the book's practical approach helps them crystallize their expectations and explain to their teams how they want certain aspects to be executed.

Students

IT students are a separate target group. Many students are interested in a career as a tester, but find it difficult to grasp what the profession actually consists of. The limited amount of time schools spend on testing is partly to blame. This book demonstrates how testing contributes to the development of an organization, and thanks to its practical approach, it also provides a clear overview of what testing actually encompasses. I also hope that the pleasure I get out of this profession – which I hope can be read between the lines – will motivate them to become a member of the great profession testing is.

Want to Know More? TestGoal.com

TestGoal is a philosophy with a practical approach, which is supported by a number of practical templates. The templates were developed together with the approach and make the application of TestGoal easier and more efficient. The templates are not included in this book. You can find them on the special Web site: www.testgoal.com.

1   Result-driven Testing

1.1   The Importance of IT

It's no longer possible to imagine life without information technology –
every day, we use computers at home and at work. Companies earn
their money by supplying IT-related services and products, or use IT
systems to run their business.

An error in an IT system can have grave consequences: a company
might not be able to supply products or bill for performed services. A
customer who is disappointed by the quality of a service or product will
look for another supplier. Software has become an entity that must be
taken seriously.

The life cycle of software has sharply declined over the past decades.
Due to the rapid succession of technological developments and changing markets, systems are constantly adapted. There are many reasons
for this, see Example 1.1.

Example 1.1: Why is software adapted? Some practical examples:

A company is launching a new cell phone. By responding to the
growing demand for advanced devices, business management expects to double the sales figures and increase the company's market share. Technically there's nothing wrong with the old model,
but it doesn't have an integrated MP3 player and video camera. In
the new model, the software for the camera and the MP3 player is
integrated with the existing telephone functionality.

D.-J. de Grood, *TestGoal*,
DOI: 10.1007/978-3-540-78828-7_1, © Collis B.V., Leiden, The Netherlands, 2008

An acquisition results in the merger of the companies' finance departments. An analysis reveals that the software that one company is using needs to be scaled up to the size of the new company. The other company has streamlined processes that can be applied to the new company with minor changes. A project is set up to implement the new processes in the invoicing system in order to create the desired synergy. Management is also expecting the invoicing system will be more efficient.

The users of a system at a medium-sized bank indicate that a number of improvements are needed. In the current system, bank statements are printed for all account holders. To save on postage, it has been decided that bank statements will only be sent to account holders when transactions have taken place. Every month, two employees go through the pile of printed bank statements to find the statements in question. The users indicate that an additional option in the application would make this activity redundant and significantly reduce work pressure.

To maintain its membership file and book catalogue, a library uses an off-the-shelf application that has been customized over the years. Although the users are very happy with the system, the supplier indicates that their version of the application will no longer be supported. The supplier advises the system administrator to migrate to the newest version so they can continue to offer help when problems occur.

The helpdesk gets a phone call from a customer about an incorrect invoice. The discount applied to the invoice does not correspond to the stated discount percentage. Closer examination reveals that the correct percentage was used in the calculation, but that the standard percentage was printed on the invoice. The number of complaints about invoices has recently increased. The helpdesk thinks that correcting the invoice will reduce the number of calls, which will reduce the waiting time, which will increase customer satisfaction.

For companies, every change made to a software program presents a risk because it can introduce new errors. And who guarantees that the business is not at risk and that the planned enhancement will deliver what has been promised?

To minimize risk, measures are taken in the software development process, for example, by developing flexible systems. Service-oriented Architecture (SOA), for example, is a system architecture that is designed to easily combine and reuse functionality. It enables new business wishes to be implemented faster [Thillard, 2006]. A second option is to choose a development method that fits the dynamics of the organization better. The ongoing trend for software products to work faster and faster means that incremental development and agile techniques are frequently chosen. A third option is to minimize risk by applying control measures, which consist, for example, of involving business in the development and assessment of the system design, and carrying out reviews and audits. Another control measure to minimize risk and increase the performance of the software is result-driven testing, the subject of this book.

1.2   A Statement about Quality

What is testing? The Dutch dictionary van Dale gives the following definition:

> A test is an assessment of the quality, capacities of people or things [van Dale].

There are a number of definitions for "test." Some of the definitions explain the term "test" more precisely. They talk about comparing the desired situation to the existing situation [IEEE 829]. Other definitions include the aim: "testing is aimed at finding the errors," in which "errors" are defined as the difference between the desired and the existing situation.

> **Error, Fault, Failure, Finding, Defect and Bug**
>
> Many terms are being used to indicate that there might be something wrong with the system. How should terms like error, fault, failure, finding, defect and bug be used?

When the analyst or the developer makes an error (or mistake) he will produce a fault. Faults are also called defects or bugs. A defect is a flaw in a component or system that can cause it to fail. Many defects hide in the code, but are never discovered. The moment they are discovered we speak of a failure, indicating that the systems does not react as we expect [ISTQB, 2007].

Findings indicate an observed difference between expected and implemented system behavior that can jeopardize the anticipated goal. This definition includes both the experience of the tester and the anticipated business goal. A finding can originate from a test fault, a fault in the test base, or a bug in the code.

Experience shows that the above definitions are accurate, but that in daily work this distinction is often neglected. Since this book is a practical book, we do the same and use "error" in most cases.

Errors are an important output of the test process. They are important because they can be solved, and hence improve the quality of the system. It must be said, however, that finding errors is not the only aim of testing. The added value of testing should not be measured by the number of findings (errors), but by the number of errors that are actually solved [Pas, 2004].

The test process has to produce information about the errors found and the degree to which the quality has been improved by solving them. In turn, the tests have to indicate whether the modified system responds to the reason that led to the change, and if it contributes to the business requirements. As shown in Example 1.1, an expectation is expressed about the extent to which

- the sales figures of the devices will double and the market share will increase
- billing in the merged company will never fail and market share will increase
- the two bank employees will be able to concentrate on their other tasks again, postage will be saved and customers will still get a bank statement
- software errors in the library package will be quickly solved without affecting the lending of books
- customers will receive correct bills and the helpdesk will receive less calls

It's not always easy to determine how a modified system will contribute to the business requirements. In the case of large and complex systems, it is particularly difficult to prove how software (a function or service) will contribute to the company's goals. This is why derived goals are defined in the development process to determine the requirements the system components must meet. In actual fact, it's a translation of the business goals to system specifications. The system design defines things such as the technical requirements for the operating system and the environment in which the system will be used.

Functional requirements define what the system has to do. Which functions are available, how does the system react when a user carries out a wrong action, which validations and business rules are applied? The system design also defines non-functional demands, such as performance and security, user friendliness and design, such as the structure and sequence of the screens or the use of a company style. The system design is a collection of documents, such as

- the requirement specifications
- the functional design
- the technical design
- UML diagrams
- the database design
- the interface specifications
- important quality attributes
- acceptance criteria
- use cases and user scenarios
- possible norms, standards, policies and legislation

If the translation is carried out correctly and each system component meets its system specifications, it is realistic to say that the anticipated business goals will be achieved. IT system testing generally focuses on checking whether the system and/or its components meet the specifications. This is called verification. Verification indicates whether the system has been "properly" built, i. e. built according to specification.

But system specifications are not always complete and clear, which is why it is important to check whether the system is usable and meets the specified goals. This is called validation. Validation determines whether the "right system" has been built, i. e. a system that contributes to the business goals. Validation prevents building a system that is technically and functionally correct but not "fit for purpose."

1.3   The Perception of Testing

Many business managers perceive IT as "too little, too late, too costly." There is a gap between business and IT on both sides of which reign incomprehension and a lack of knowledge about the other discipline [Ommeren, 2006]. This is why business management often has wrong expectations of software development and testing.

A software development project consists of building applications and systems, making software development an activity that produces a product. Business may perceive the product as being too little, too late and too costly, but in the end they have a system. Many people actually believe that testing is an activity that does not produce anything at all. The system will be built anyway, tested or not.

The testing community has difficulty explaining what the added value of testing is to management, customers and developers [Clermont, 2006]. As a result, testing is frequently pushed into a subordinate role and cannot fully contribute to the development process.

Provided testing is carried out well, it makes a positive contribution to the software development process. Testing adds value to the development process by

- Contributing to the development of "fit-for-purpose" systems that work according to the requirements, quality attributes and expectations (implicit or other). It achieves the goal.
- Preventing damage while the system is live because errors will have been detected during testing and solved on time. This takes away the cause.
- Preventing damage while the system is live because the errors are known and their consequences have been anticipated. This reduces the impact.
- Providing insight into the quality of the test object, which instills confidence in the test object.
- Providing insight into the quality of the test object and its progress, making effective project coordination possible [Black, 2002].

It is important that business management show commitment and enable testers to do their job professionally. To get this commitment, testers will have to communicate with management in a language they understand. Testers have to realize business managers do not think in functions, but in products and services; not in incidents, but in wishes and requirements. Business managers like hearing which wishes and requirements are feasible. They prefer hearing about the risks that are being dealt with than about the problems that still exist. In short, a manager's involvement is obtained by giving them the information they ask for, not data.

Business managers will also become increasingly involved if they have insight into the testers' activities. They are involved when crucial decisions need to be made to reassure them that testing will answer their questions, and that only the activities required to answer those questions will be carried out. This requires a specific view on testing, which we call result-driven testing.

Business managers are not the only people who have a stake in the system that is being developed. Result-driven testing also involves a number of other stakeholders.

1.4   A Common Goal

Software is changed to eliminate the cause of an error, or because the users have additional wishes. New requirements sometimes lead to the replacement of existing applications. Key notions in this text are the elimination of errors, additional wishes and new requirements. These notions imply that each change in the software is related to a stakeholder that is having problems with an error, would like to see his wishes implemented, or has come up with new requirements. In addition to business managers, the following groups are involved:

- **Users**
  The people who use the system. Users can be end-users, for example, customers who place orders at a Web store, or company staff, such as order processing clerks or managers who need sales reports.

- **System Administrators**
  The people who are responsible for the tasks and activities that are required to maintain the system in such a state that it continues to run according to the known requirements and needs [ASL].

- **Controller**
  The person who carries out checks and provides insight into the company's operations. Many companies must provide proof of the transactions they perform, the processes they run, or the quality measures they implement. The reason behind this may be a service

level agreement that includes traceability. Government institutions also demand that a company's operations are "auditable." Examples are the requirements defined by the Internal Revenue Service, the Sarbanes-Oxley Act, Basel II and the Tabaksblat Code (Dutch Corporate Governance Code) [van Es et al, 2005].

All of these stakeholders have their own job-related wishes and requirements. For example, users will want a fast and user-friendly system, while the system administrator will be more interested in a system that is easy to maintain. The controller will find traceability (the degree to and the ease with which the correctness, completeness and integrity of information can be checked) an important quality aspect.

Testing is part of a chain of activities that are required to change or develop software. Testing is usually part of a project, which has a project organization. In this project organization, the tester works with various parties, namely:

- **Developers**
  Changes and enhancements require changes to be made to the program code. This is the job of the developers. Ideally, developers base their work on a design created by analysts. If there is no design, the developers decide what is done and how.

- **Analysts and System Designers**
  The analysts and system designers translate the wishes and requirements of the business, the users, the system administrators and the controllers into a system design, which defines how the system should work and how the code needs to be changed.

- **IT Architects**
  IT architects specify the framework within which the system will be developed and the long-term guidelines the design must meet. Because IT architects like their systems to be built within the specified framework, they are considered to be quality driven [Dietz].

- **Project Managers**
  The project manager's task consists primarily of setting up the development project. Project managers ensure that the intended solution is realized. Project managers direct and facilitate, which does not release them from having technical knowledge, because specific knowledge is needed to make the right decision. Larger projects often distinguish between technical and general project managers. General project managers rely on the knowledge of their technical project leaders, as they have less in-depth knowledge of the project's activities.

- **Customers**

  Customers are people who have test budgets and commission test activities. The customers' hierarchical position makes them an important stakeholder with whom testers communicate intensively. Customers are usually among the group of previously described stakeholders, i.e. business managers, project managers or test managers .

Each of the mentioned stakeholders have their own job-related issues and interests. Their daily work determines the issues they focus on. Each party will also try to leave their mark on the system.

Fig. 1.1 A software development project involves a number of disciplines, each of which has their own issues and interests

Interests that diverge too much during a software development project jeopardize the success of the project. Conversely, the following can be said:

> An important success factor for a project is the degree to which the disciplines involved pursue the same goal .

If the stakeholders intend to pursue a common goal, which goal should it be? The answer is simple. It has to be the goal the organization is pursuing; the business goal.

In a result-driven project, testers, developers, project managers, system administrators, users and other stakeholders all work toward the same goal. Ideally, every activity that is carried out contributes to achieving the common goal. We could speak of result-driven development, result-driven project management, result-driven system administration, result-driven work, and of course ... result-driven testing.

1.5   Tying in with the Business

Focusing our activities on the business goals should enable us to bridge the gap with the business managers. It is crucial that we understand the goals the business managers are aiming for and that the business managers understand our point of view.

Let us listen in on the board meeting of a fictitious company. The business managers are talking about conquering new markets and increasing sales revenue. "How's our market share doing? Can we reduce costs by streamlining operational processes?" The company has legal obligations, and politics also play a role. One manager has a hobbyhorse and is ready to sacrifice everything to make it a success.

The managers know that most of their plans will involve IT. That may be why they discuss software development projects now and again. Testing, however, never makes it on to the agenda. They're not IT driven: software is no more than a means of achieving the defined business goals.

Many organizations use business cases to drive their projects because they justify the investment. A business case describes the business goals, the advantages the project has for the organization and the degree to which the project contributes to the business goals. The core of the business case is the assessment of the pros and the cons. The project's goal is assessed against the required effort [PRINCE2, 2005].

Despite the fact that testing is not on any of the meetings' agendas, result-driven testing is discussed because it is at these meetings that the business requirements and wishes are determined. The test project assesses the suitability of the developed products and the degree to which they fulfill the business requirements and wishes. In this assessment, one has to look further than the financial result of the business case. The non-financial results and the implicit expectations are also part of the anticipated goal.

Example 1.2

Example 1.1 shows a number of reasons why software is adapted. The company that is launching a new cell phone expects to double its sales. The financial result is quite easily derived. The profit gained from additional sales has to balance with the device's development costs. But even if the numbers balance, it doesn't necessarily mean that the company's market share has increased. The new device may have sold well, but the competitors didn't sleep. The device that the competitor launched was also well received by consumers. As a result, the market share of the company in the example did not increase and the anticipated goal was not achieved.

Throughout the software development project, the test project provides information about the quality of the system being built. Testers would be well-advised to compile this information in a language that business managers understand. Incorporating the anticipated goal as frame of reference in the information gives managers an idea of the progress and the bottlenecks encountered on the path to achieving their goals.

1.6   Result-driven Testing

Result-driven testing assesses the quality and suitability of the developed products as well as the extent to which they meet the business requirements and wishes. In order to do this well and efficiently, we must

- know and understand the anticipated goal
- only carry out those activities that contribute to the anticipated goal, or that produce information about the extent to which the goal has been achieved
- provide the stakeholders with understandable information in a timely manner

Result-driven testing is therefore defined as follows:

Result-driven testing puts the business goals at the center of each step in the test project. The project focuses on providing added value for the anticipated business goals and for the software development project.

Result-driven testing puts the business goals at the center of each step in the test project. The reason for the development or modification of the system is defined at the beginning of the test project. While creating the test assignment, the tester not only establishes what the anticipated goal of his assignment will be, but also what the business wishes are. The difference between the two becomes clear in the following example.

Example 1.3: The Connecta project

A test coordinator is discussing a new test project with his customer and asks what is expected of him. The customer replies that he wants the system to be properly tested. A detailed system design has been created and the customer wants to know if the system can meet the requirements. During the test, the test coordinator must adhere to both the company's rules and the test's guidelines. The customer wishes to receive regular information on the progress, status and risks encountered during the project.

This may sound like a clear assignment, but it does not say much about when the test coordinator can give a positive release advice. Doing this requires information on the goal the company is aiming for. Asking about the crucial success factors reveals that the company wants to streamline its activities. The new system is supposed to replace a number of existing systems. Transferring the data manually between the various systems is too time consuming and prone to errors. The customer expects the new system will help prevent many of the errors and unnecessary delays. The investment in the new system will be recovered because fewer FTEs will be required to carry out the activities.

The test coordinator thanks the customer for this information. He now knows which questions he will have to answer. He sets up the test project so that he can give a well-founded release advice that answers the questions.

Result Aimed and Result Driven

Result aimed: We aim to achieve the result in everything we do. Result driven: We are driven by the result, so we only do things that have added value. The added value is determined by the extent to which the activity contributes to the anticipated goal or produces the desired information about the extent to which the goal has been achieved. We focus on this throughout the duration of the activity (the same applies to "result aimed").

1.7   Focus on the Goal

In the previous sections, we learned about goal. Now that we understand what they are, we can use this knowledge and this notion to put the goal at the center of our test project. The anticipated goal is used to

- describe the test strategy
- sets up the tests
- build bridges to the organization
- run tests
- generate test reports
- write a release advice

The anticipated goal determines the test strategy. The test coordinator uses the information he gathered during the assessment of the anticipated goal to draft a suitable test strategy. The purpose of the test strategy is to establish as effectively as possible whether the anticipated goal will be achieved. In Example 1.3, the system should, of course, function well, but the real question that has to be answered is: does the new system save resources now that the processing has become faster and contains fewer errors?

The goal is also central during the set up of the tests. First, a sanity check is run to determine whether the system documentation supports the test strategy. The test base is also tested against the anticipated goal to obtain a timely indication of whether the system design will produce the right system. The goal determines the structure of the test design, which ensures that all of the test results can be related to the identified risks, i. e. the areas that need attention. This bridges the gap between the world of the business managers and the other stakeholders.

While setting up the test design and running the tests, the testers will build bridges to other parts of the organization. Because everyone is aiming for the same goal, the testers and other stakeholders speak the same language and work hand in hand. The testers never lose sight of events that may arise from risks, requirements or the market, and they can indicate how the test approach needs to be adapted to respond to them.

During the test, the goal is taken into account by running the tests that provide insight into the quality of the system the quickest. The tester assesses the test object against the anticipated system behavior (described in the test design) without losing sight of the fitness for purpose. The tester shares the information produced by the tests with all of the stakeholders.

The anticipated goal is defined according to a number of Key Performance Indicators (KPIs). Throughout the test project, the test report (see Chap. 20 Test Reports) provides insight into the status of the KPIs. Reporting on the same KPIs from the beginning produces clear information about the progress of the test and its results, and prevents surprises arising after the test has been completed.

The final test report is produced at the end of the test. This report provides insight into the activities that were carried out during the test. It refers to the test strategy to indicate how the test was run, and builds a bridge to the anticipated business goal. This enables the tester to give a release advice that is in line with the business managers' perception, and to explain how he reached his conclusion.

2 TestGoal and the Ten Test Principles

2.1 Test Principles

Testing is commonly carried out according to a test methodology. The test methodology ensures that testers and other stakeholders use the same terminology and definitions. The methodology also contains best practices so that the knowledge and experience of others can be used. This contributes to the efficiency and quality of the test project.

But there's more to testing than just the methodology. Many organizations apply a methodology but not all of them implement it properly. There are testers who do everything according to the book but can't find their feet in the organization. After all, strict adherence to a specific

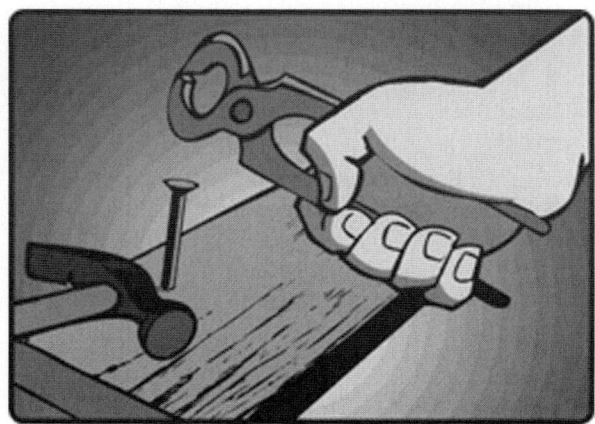

Fig. 2.1 It is not the method, but the way it is applied that determines the success of a project.

D.-J. de Grood, *TestGoal*,
DOI: 10.1007/978-3-540-78828-7_2, © Collis B.V., Leiden, The Netherlands, 2008

methodology is no guarantee for success: it is not the method, but the way it is applied that determines the success of a project

The correct application of the methodology depends on the mindset and the expertise of the tester, who has to be able to stress the right things and make the right choices. Ten test principles have been developed to help result-driven testers put this into practice. The test principles are a short and strong formulation of the tester's desired mindset, knowledge and working method. They indicate how the tester can create added value and how he can make it visible to the organization.

The test principles enable testers to focus on the anticipated goal without external assistance. By thinking and acting according to the ten test principles, testers assume a result-driven mindset. They make sure that they apply the test methodology in such a way that it achieves optimal added value for the organization.

The result-driven tester works according to the following principles.

Table 2.1 The ten principles of the tester

	Focus on result		Test in phases
	Build trust		Facilitate the entire IT life cycle
	Take responsibility		Provide overview and insight
	Master the testing profession		Ensure reusability
	Build bridges		Keep in mind: Testing is fun!

2.2   Focus on Result

Result-driven testers will always focus on the business goal, regardless of what they're doing. Their drive extends far beyond contributing to the quality of the software. The tests they run must be able to reveal

whether the system helps to reach the goals that are being pursued. Reaching the goals will contribute to the companies business result. In this context the companies result could be a non-financial one or aimed revenue

Result-driven testers do more than just verify, i. e. demonstrate that the system is built according to the system design. They also validate, i. e. provide insight into the extent to which the system fits the business processes and whether the business goals will be achieved.

Working from the organization's anticipated goal enables testers to look after different interests. For example, a functionally correct solution could be very user unfriendly or an application that is easy to maintain could conflict with the security requirements. Result-driven testers not only inform stakeholders when their interests are at risk, they often have better solutions.

Focusing on the goal also prevents investing excessive energy in relatively unimportant matters. Result-driven testers are critical about their activities and continually ask themselves how the activities fit the organization and contribute to the goal. The testers' result-driven mindset adds value to the organization. The organization knows that the tester will look further than the tip of his nose.

2.3   Build Trust

Confidence in the system can be built through intensive testing. Intensive testing reduces the risk of errors being overlooked, but is unfortunately not always feasible; there is rarely enough time to test every situation. This is why risk-based testing is used, and the most important parts of the system tested more thoroughly than others.

It's easy to find fault in people's work. Good testers not only look for errors in other people's work, they also look for the good things. By relating the test goals to the business needs and requirements, business management can establish the areas that need attention and the components that work well. As testing progresses, confidence in the system grows. By indicating which parts of the anticipated goal will be achieved, testers show that they understand the business goal and are pursuing the common goal.

Testers make sure they can be trusted and that a positive release advice can be seen as a guarantee for quality software. This also works the other way around; when testers indicate that the quality is poor, they expect their indication to be taken seriously and their recommendation to be followed. Testers can build trust in a number of ways. First, by doing a good job, which is the main prerequisite to building trust. They also have to carry out the right activities, and those are defined in consultation with the stakeholders.

This is why the stakeholders should be actively involved in the determination of the risks and the areas that need attention. Stakeholders are often concerned about one specific component or anticipate certain problems. If the tester takes things seriously, the stakeholders will take the tester seriously. The tester shows which tests he is going to run to determine whether the concerns are justified and explains how he uses his test expertise to make sure the anticipated problems do not occur in the live system.

Trust is also a result of thoroughness and transparency. Do not spend time and energy on hiding mistakes. Everyone makes mistakes, even the test team. People in the organization will start worrying if they have the feeling that there are hidden problems. It's never pleasant when a tester has to admit he has a problem or has made a mistake. Experience, however, shows that others appreciate their honesty and are often prepared to accept the consequences. A good solution is often found together.

Example 2.1: Trust

The functional acceptance test for a new software development project is almost finished. A lot of tests were run over the last couple of weeks, and there's one week to go. A concern emerges during a meeting with the maintenance organisation: all of the live data has to be migrated to the new database. The system administrator knows that tests have been developed for this, but he recently heard about a situation that rarely occurs and is now wondering if it will also be tested. The test coordinator is starting to feel insecure. Should he say that the migration tests have been discussed in detail and that he is confident that this situation has also been tested? Or should he admit that he did not anticipate this situation at all. But if additional tests need to be run, the planned tests will never be finished on time.

The system administrator is really worried about this issue. Faultless migration was high on the list of business goals. The test coordinator decides to give this issue additional attention. He and the customer investigate how the situation can be solved and quickly find the answer. They decide that one of the testers will look into the situation with a system administrator and will amend the test set accordingly. The activities that were planned for the coming week, the testing of low-risk functions, have already been documented. The project manager asks a programmer to run the tests, which is not the ideal solution because a programmer is not a tester, but the test manager knows the programmer and knows he can do it. "And if he works with us in the testing room, I can keep an eye on the situation," he thinks.

The migration tests, which were initially a showstopper, are ready on time. The programmer also found a number of things in his colleagues' code and enjoyed seeing how the test team actually works.

2.4   Take Responsibility

At the end of the test project, the tester gives a release advice that he would "bet his booty on." The tester ensures that he has prepared the right tests and that he has taken the risks and the necessary coverage into account. He keeps an eye on things during the test project and makes sure that all of the activities that have to be carried out actually take place. As a result, he can vouch for his advice and support it, even if it's negative.

Testers represent the test domain and are also strongly involved in the rest of the product life cycle. A lot of the problems that are discovered during testing were created somewhere else in the development process. Below are a few examples.

- **Bad Assessment of the Customer's Wishes**
  The customer's wishes were not properly assessed. However excellent the programmer is, the solution will never achieve the anticipated goal.

- **Poor Design**
  The design was not properly thought through and is not detailed enough, or it contains contradictions and gaps. When a design is

faulty, the developer needs to fill in the gaps. As a result, the different components do not always interact well.

- **Poor Programming**
  The quality is poor. The code contains at lot of bugs because there wasn't enough time to write neat code, or because the programmer is inexperienced.

- **Poor Configuration and Version Management**
  The development team is working hard to solve errors. As a result of poor configuration and version management, errors are being solved in the wrong code or wrong software components are included in the release.

A responsible tester shows stakeholders where the bottlenecks are and explains how they affect the anticipated goal. If necessary, the tester offers help and expertise. The tester also draws the stakeholders' attention to issues that still need to be taken care of.

Taking responsibility also means that the tester does more than just test. He sometimes can't avoid taking on tasks that are not his responsibility, such as release and configuration management. Because these processes are crucial to achieving the anticipated goal, the tester here too takes responsibility if necessary.

Sometimes the goal is best achieved if the tester relaxes his requirements. Nobody benefits from sticking to requirements that the tester knows will never be met. It is much better to adapt the strategy, within the realms of possibility, in order to obtain the best possible result. Clear boundaries must be set though, because too many compromises can be counterproductive.

2.5   Master the Testing Profession

A good tester knows that testing information systems is not just a matter of pressing a few buttons. He masters the test profession, which is aimed at giving an organization the confidence that an information system works properly. This means that a good tester has to have sound knowledge of information technology, system development, test methodologies and the processes in an organization. His knowledge of the different business areas also enables him to understand the test object and to be an interlocutor for the users and the business managers. This enables him to build bridges to all of the stakeholders.

It goes without saying that a tester is familiar with accepted testing standards, methodologies and techniques. This knowledge and experience enables him to select the elements that best suit the purpose of the test. Each test project is unique and it is up to the tester and his skills to define a suitable and efficient approach.

Example 2.2: An inexperienced organization

A tester is asked to set up tests for an organization that has little experience building and testing software. The customer explains that they have subcontracted the development of the system, but that they want to run an acceptance test upon delivery. To prepare and run the tests, three of the organization's employees are assigned to the tester. The users have never tested software before, and are looking forward to the challenge.

The tester realizes that the organization's resources are limited. He decides to let the users do what they are best at. A number of business scenarios will be explored during the acceptance test. The tester explains the basic principles of testing and how he is going to approach the acceptance test. He expects the users will gradually gain a better understanding of both testing and the system. He therefore introduces elements of "exploratory" testing . This enables the users to adapt the predefined scenarios according to their experience and even add new ones.

In order to test the functionality and the business scenarios, the tester turns to the contractor. The contract states that he has to supply tested software. The tester shares his requirements for the test plan and test design with the contractor, who reviews the test plan and checks the coverage of the test design. The contractor is cooperative and adopts the suggestions for additional test cases without resistance.

The tester discusses the approach with his customer, who agrees. Both test projects should provide good coverage. The internal working of the system has to be tested as well as its applicability.

A good test requires thorough preparation. Test plans, test specifications and test scripts must be feasible and clear. It must be possible to relate the test results to the business wishes and requirements. Thorough preparation, however, does not mean that testers cannot be creative. The rapidly developing "agile" development and test methods actually encourage creativity.

In addition to general subject knowledge, testers often have a specialism such as test automation, performance testing or security testing (see Chap. 3 Test Expertise).

2.6   Build Bridges

Result-driven testers are the central point of contact. They communicate with all of the parties involved in the software development project. They know who has a stake in the IT solution and, if necessary, they mediate between the parties.

Testers have contact with the accepters to stay in touch with their requirements and wishes. Often a project will have to deal with changing requirements, risks and expectations. The contact with the accepters enables testers to anticipate developments in and changes to technology and business.

Because testers can put themselves in the stakeholders' position, they can explain to developers how the stakeholders perceive the system. Testers have enough technical knowledge to understand developers and help them analyze the errors or the system's behavior. If necessary, they will bring developers and users in contact.

> *Example 2.3: Changes*
>
> Changes to existing systems are implemented in the company on the basis of change requests, i. e. "changes." Changes have to be tested before they can be implemented. Based on the impact of the change, the tester will work with a system administrators or user.
>
> A user is involved in testing change CR0001745. The user has reported nine errors, which is quite a lot for a small change. The tester discusses the change with the project manager, who says that everything's fine. "I immediately assigned a developer to it," he says, "and he says there's just one real error; you'll find the fix in the test environment tomorrow."

It is not rare for different stakeholders to have different ideas about the nature and the importance of an error. The tester knows that a number of errors were found. His guess is that the errors are more annoying for the user than they are difficult to fix. He intervenes. The tester explains to the developer that there is a reason why the user logged nine errors, and that he probably won't be very happy to hear that the builder doesn't find his errors important enough to fix. The tester suggests the builder and the user talk to each other and decide together which errors will be fixed. In the end, the tester was right and the remaining errors were indeed easy to fix. The fix is available for retesting the following day. While discussing one of the errors, it turned out that the error was not caused by the code but by the specifications. It was agreed that the tester discuss the issue with the analyst.

Result-driven testers also build bridges between the project organization and the line organization. A lot of the knowledge that is gained from a software development project is also valuable for the line organization, for example, knowledge the test team gained about the configuration of the system. It's also possible that test data was used for the configuration of the test project. The errors the test team made don't have to be repeated by the maintenance organization. The test design is handed over so the maintenance organization can carry out the necessary tests when future changes are made.

The line organization often has a checklist that has to be worked through before the project can be formally completed. If a checklist hasn't been made yet, the tester can suggest making one. Discussing the expectations of the accepters in advance prevents surprises arising. This is the tester's contribution toward a smooth transition from the project phase to the live phase.

2.7   Test in Phases

 Experience shows that phased testing produces the best quality because the result-driven tester is using a clear project approach and planning his activities in phases. A software development project has a number of test projects. A project is created by specifying its goal. This ensures that tests are not carried out twice or forgotten. A test project can consist of several test phases. The work breakdown structure (WBS) and planning (see Chap. 9 Test Budgeting and Planning), establish which

activities have to be carried out when. This ensures that the activities are carried out in a logical sequence and that the most important risks are covered first.

A phased test project has the following characteristics:

- The strategy fits the software development process.
- A structured test methodology is used.
- If the waterfall model is used, the components are tested first, then the integration of the components, and finally the link to the business processes.
- If the development and delivery are done incrementally, the tester first discusses the delivery schedule with the build coordinator. He then ensures that the increments are delivered in a sequence that minimizes the biggest risks first.
- The quality attributes are distributed over different test levels . The tester schedules time to test the functionality, performance, interoperability, user friendliness and security.

A phased test process has a number of advantages:

- The different stakeholders can do their preparatory work. While the developers work on the first delivery, the test team makes sure the corresponding test designs are finished on time. The analysts are familiar with the content of the next delivery and submit the amended system specifications for review.
- Entry and exit criteria can be used. These concrete agreements about the phase transitions clarify the expectations. The criteria also help create a clear progress report, and they prevent later phases suffering from oversights in previous phases.
- Expectations can be clearly expressed. Step by step, it becomes clear whether the system being tested will meet the expectations. The further the test progress the clearer it becomes whether a system can go live.

2.8   Facilitate the Entire Product Life Cycle

Testing is an integral part of the IT development chain and life cycle, and makes demands on preceding links in the chain. The added value of a tester is determined by the extent to which he can influence the chain to ensure that the final information system comes as close as possible to the business requirements and wishes. A good tester is therefore involved in the entire life cycle, reminds others of their responsibilities and controls the preceding links in the chain.

The tester can benefit from demonstrating his involvement early in the process. Doing this enables the tester to help during the analysis and design phase by testing the products for their testability. He can also test whether the design indicates how the anticipated goal is going to be achieved. For example, the organization in Example 1.3 expects that the live system will be free of errors and delays. The tester can check whether the system design specifies how this is achieved.

The tester encourages the developers to test their work during the development process so errors can be detected as early as possible. He explains the importance of module testing and, if necessary, suggests how it can be approached. In terms of his own activities, he knows that the designed tests will not lose their validity at the end of the test project and hence ensures they can be reused.

After the tests have been run, the tester shares his knowledge and testware with the maintenance organization. A good handover contributes to the smooth transition from project phase to maintenance phase.

2.9   Provide Overview and Insight

 Many different kinds of data come together in a test project. To start with, a system has several functions, each of which is submitted to a large number of tests. Test results not only refer to the function they test, they also cover risks, issues and quality aspects. There are also errors and the test project has its own progress and bottlenecks.

In order to turn all this data into useful information, the tester must be able to see the main features (overview) and explain the details (insight). The tester provides overview and insight by producing a clear test report in which he specifies the quality of the test object and the progress of the test project. The tester uses clear language to describe the consequences for the anticipated goal. This is how he contributes to the insight that is required to choose the right solution.

By using a transparent working method, the tester builds trust and makes the information accessible to all of the stakeholders. This prevents confusion and unnecessary communication and thus helps achieve the common goal.

Example 2.4: An Intranet site for the information

Following several confusing reports about the progress of the test project, the test coordinator decides to make important information centrally available. A colleague has shown him how easy it is to save Microsoft Word documents as HTML files. The test coordinator puts the information he wants to make centrally available on the Intranet.

Test Reports
The test report is updated every other day. There's also a weekly meeting, but people often ask what the project's status is and how it's progressing between meetings. They can now check the current status at their desk.

Planning and Milestones
To achieve the goals defined in the planning, the stakeholders must be aware of the planning. In the past, the test coordinator discussed the planning with the customer. The planning was sometimes adjusted during these meetings, but the testers didn't always notice. The team now works with one planning and has insight into the milestones it needs to achieve.

Overview of Test Environments
The test team uses a number of different test environments, each with its own version, configuration and login details. Placing this information on the Intranet prevents the question "What's the login for the test environment again?"

Responsibilities
The test coordinator creates a list of the employees and their specializations and responsibilities. In this way, the developers and the users always know who they can address for specific issues.

Everyone is pleased with the approach and the test coordinator notices that the testers in his team contribute even more information. The project manager asks if the agreements from the error-meeting can be added as well. "Everyone will know where to find them," he adds.

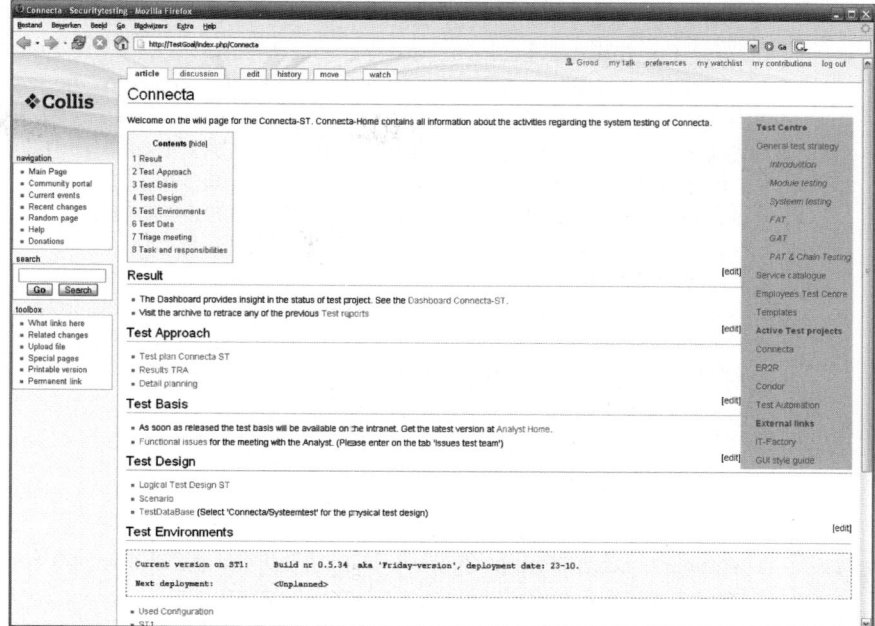

Fig. 2.2 The Intranet site in the example has indeed grown. Screenshot of a test project page

2.10 Ensure Reusability

Producing products that are reusable for subsequent test projects contributes to their efficiency. The result-driven tester knows that the tests he uses will not lose their validity at the end of a test project. While using the system, users will invariably discover errors that need correcting or will want to make changes. The organization therefore needs a reusable test set. Tests that are expected to be run frequently can be automated.

Reusability is also important for another reason. A good test project takes time and stakeholders see testing as a costly activity. Good reusability ensures that the time that is invested in the preparation and building of knowledge will prove useful even after the test project has been completed. This benefits the business goals in the long term.

The tester has a number of options to increase the reusability of the products. To ensure that the purpose of each product is clear, they should have good descriptions and instructions.

A few examples: Adding comments to the test scripts helps understand the test's goal. A physical test case specifies which part of the test base is covered. As a result, a successor can easily check whether the tests still apply to the last version of the specifications. The risk analysis not only takes the relative importance of the functions into account, but also the people involved in the risk estimation. This enables a successor to convene stakeholders when the risk analysis needs to be updated.

But good descriptions and instructions are not enough: you have to be able to find the products. The result-driven tester also spends time on the handover so that people who need to work with the products at a later stage know they exist and where to find them.

2.11   Keep in Mind: Testing is Fun

 Testing is a fascinating profession. Happy and enthusiastic testers convince others more quickly and work more efficiently.

Testers see a lot of suboptimal processes and faulty systems – this can cause them to adopt a negative attitude. But it won't be the first time that a delivered system doesn't work, and it won't be the last time that a bad design was used or the tester wasn't given enough time to test.

This doesn't change the fact that testing is a wonderful and fun profession. Result-driven testers are aware of their position and motives. They know they add value to the software development project.

Testers share the responsibility that systems only go live if they are of good enough, i. e. they meet the business requirements. The errors the tester finds not only improve the quality of the system, it also guarantees they do not occur in the live system. A tester contributes to the goal, and that puts the tester in a unique position. He is one of the first parties to gain insight into how the system works and has contact with many of the stakeholders in different disciplines. The production of a high-quality and viable IT system is an ongoing challenge.

Testers have to be flexible and creative and able to find a suitable solution for every new situation. Testers develop continuously and in many directions. Testers grow by taking on more responsibility and acquiring methodical knowledge as well as by learning special technical skills such as test automation, and performance or security testing. All this makes testing a fascinating and versatile profession.

2.12   Applying the Test Principles

The following needs to be said about the test principles:

The ten test principles are not autonomous; they are correlated. The stakeholders' active involvement in determining risk areas increases the confidence in the test approach. As soon as the tester approaches the stakeholders, he builds a bridge. A clear project phasing not only contributes to the "test in phases" principle, it also provides "overview and insight." Together, the test principles are the foundation for result-driven testing.

A number of descriptions are provided with an example to add color to the principles. To show how the test principles are applied in practice, they are regularly referred to throughout the rest of the book. A pictogram in the margin indicates that a test principle is actively applied.

The reader is challenged to use the test principles and to apply them in their own way. The principles do not necessarily have to be followed as they are described in this book. It is important, though, that the principles become part of the tester's working method. A tester who does not apply the principles is doing neither himself nor his customer a favor. It is therefore necessary to regularly check that all of the testers are paying enough attention to all of the test principles; this keeps them on their toes and focused on providing added value.

3   Test Expertise

The test principles are a guideline for the motivation and mindset the tester has to assume in order to work in a result-driven way. They indicate how he can add value by testing. A structured test methodology is also indispensable. Together, the test principles and the test methodology provide the starting points for the set up and execution of a test project. But there is also a third aspect that is very important for the result-driven tester. However complete the methodology is, it can never cover every situation. There will always be choices that have to be made based on the tester's knowledge and experience. A tester's expertise is therefore of great importance.

Testers should have knowledge of IT, as well as of the industry they work in. To be able to build bridges to others and communicate with them efficiently, they also have to understand other professionals in the organization. This requires flexibility, because at one time the tester can be having a detailed discussion with a programmer or analyst, and at the next trying to understand the users' requirements by putting himself in their shoes. To provide overview and insight and to explain the release advice to management, the tester must also be able to distance himself from an issue.

And last but not least, the tester is familiar with the testing standards, methodologies and techniques. He is able to combine the best elements from these methodologies, which enables him to find and manage errors.

The following paragraphs describe the test expertise that is required for a number of roles. These roles are not always clearly distinguished in every organization. In some projects, all of the roles are fulfilled by one tester, who combines the roles of knowledge expert and test coor-

D.-J. de Grood, *TestGoal*,
DOI: 10.1007/978-3-540-78828-7_3, © Collis B.V., Leiden, The Netherlands, 2008

dinator. In larger projects, the roles are generally separated. To keep the test approach generic, this book mainly speaks about the tester. Testers can have one or more of the roles described below.

3.1   The Test Manager

If several test projects are carried out in a software development project, the test manager is responsible for the overall outcome. He decides how the master test process is approached and ensures that the combined test projects have sufficient test coverage. It is important that the coverage is complete and that things are not unnecessarily tested twice. He also monitors the progress of the individual test projects and guides the test coordinators. The test manager also manages the generic test strategy and master test plan (MTP) if they exist.

In order to do this task well, the test manager has to have a good understanding of the business requirements and wishes. He also has to have project management skills *and* experience as a tester. The test manager is involved in the whole development process. He is also active during the goal, approach and assurance steps (see the step plan in Sect. 4.3.1 for an explanation of these steps).

3.2   The Test Coordinator

The test coordinator defines the test approach that is going to be used and is responsible for the success of a test project. To achieve the anticipated goal as efficiently as possible, he determines, for example, whether he is going to work out the details of his physical test cases or use an agile test approach, and whether he is going to run his tests manually or automatically. The test coordinator works according to the guidelines in the generic test strategy or the MTP. If these documents are not available, he consults the stakeholders and determines his course. He describes the test approach in the detailed test plan (DTP). In addition to drafting the DTP, the test coordinator also leads one or more testers. He monitors the progress and the quality of the test activities and discusses the test results with the people involved. He is an important player in the error-management procedure. In the end, it's the test coordinator who decides whether a positive release advice can be given.

In order to do this task well, the test coordinator has to be an experienced tester *and* must have project management skills. The test coordinator determines and monitors the day's activities and is therefore active during the entire test project. He is also active during the start-up phase and the goal, approach and assurance steps.

3.3 The Test Analyst

The test analyst uses the test plan to determine which tests are carried out. He understands the anticipated goal and examines the test object's areas of attention. The test analyst masters his test design techniques and knows when to apply them. Because he communicates a lot with the analysts and designers, and because system design is his area of expertise, he knows about system design techniques. For example, he can read UML diagrams and he knows which requirements a good use case has to meet. The test analyst is mainly involved in the design step and creates the logical test design.

3.4 The Test Engineer

The test engineer is responsible for the physical test design and the execution of the tests. He elaborates the tests. If test analysts have been appointed to create the logical design, the test engineer translates it into detailed test scripts. If there is no logical test design, the test analyst will use the test base as his foundation. Test scripts are not created in agile environments. In this case, the emphasis of the work shifts from the design to the test execution, since the test engineer is also the one who runs the tests. The test engineer also records and discusses the errors. He not only builds bridges to the designers and programmers, but also to the test coordinator. The test engineer is primarily involved in the design, set up and execution steps.

3.5 The Test Specialist

Testing also has a number of technical specializations, such as test automation, performance testing and security testing. The testers working in these areas are expected to have additional technical expertise.

- **The Test Automation Developer**
  The test automation developer is the test team's automation expert. He can be involved in the creation of automation projects, tool selections and pilots. The test automation developer is the programmer among the testers. He implements the logical or physical test designs in test scripts so the tests can be run automatically.

  The test automation developer has testing and programming knowledge. His knowledge of tooling enables him to program and parameterize test scripts. He uses his technical knowledge of the system to configure the test environment and include test data in his test scripts. The test automation developer is primarily involved in the set up and execution steps.

- **The Performance Tester**
  The performance tester checks whether the system meets the load, stress and reliability requirements. These tests distinguish themselves from functional tests by their highly technical nature. Because simulators and load generators are used in performance testing, performance testing overlaps with test automation. A lot of measurements and calculations are made during performance tests. Error analysis and significance are issues that play a role. Once the performance has been determined, it often has to be improved.

  Because of his technical knowledge, the performance tester is able to determine the system parameters that influence the performance together with the system specialists. If necessary, he changes these parameters to improve the performance. To this end, the performance tester has knowledge of IT architecture, database organization and the systems' workings. The performance tester is involved in the design, set up and performance steps.

- **The Security Tester**
  The security tester determines whether the system meets the defined security requirements and checks them against the organization's policies and procedure. To this end, he performs reviews and audits on the design, the code, and the procedures, and carries out structured security tests on the system. He also assesses the test results. He is able to recognize possible security risks on the basis of errors or combinations of errors.

  The security tester distinguishes himself from the other testers because he has up-to-date knowledge of the most important security exploits, which enables him to quickly recognize the most common security errors. The security tester can familiarize himself with the common pitfalls that come with, for example, the used design methods, design platforms, or middleware. He has knowledge of the test design techniques , which he uses to analyze the network or system in a structured manner. The security tester can be involved in all phases of the test project. He performs a review during the sanity check and checks the system while the test is running. In the meantime, he can audit the code or the architecture.

The testers' expertise and quality are crucial for the success of a test project. Another important element is the way in which the test project is set up, as it determines whether the tester's expertise is deployed properly and effectively.

4 The Approach

Result-driven testing can be applied to any test method. The previous chapters explained how expertise and the TestGoal principles ensure the methodology is correctly applied. The ten test principles help testers create a result-driven mindset. Result-aimed thinking is an important step for the tester, but it would also be nice if the used methodology supported result-driven activities. TestGoal was developed according to the idea that the testers' expertise, the testing principles *and* the methodology have to be integrated and have to support each other. In Test-Goal, result-aimed thinking and result-driven acting form a unity, which creates a practical test approach that puts the business goals first.

The TestGoal approach helps testers use their expertise and knowledge and apply the test principles. A clear step plan shows when activities have to be carried out, how they can be carried out and which areas need attention. Where necessary, the approach also provides clear information about best practices. The approach gives the test project a clear structure, which ensures that even the less obvious aspects are not overlooked. Moreover, communication between project members is more efficient and the anticipated goals are known beforehand. As a result, the activities get off to a good start and trust is built from the very beginning.

The TestGoal approach can be applied to a wide range of test projects. The approach is fully autonomous, but can also be used in combination with other methodologies and standards, such as TMap [Pol et al, 1999], [Koomen et al, 2007], ISTQB/ISEB [Spillner et al, 2003], the IEEE standards for software testing [IEEE 829, 1998], ISO 9126 and BCS SIGIST [BCS SIGIST, 1997]. Choosing one option does not exclude another. The TestGoal step plan is a guideline for the activities that are to be carried out. The applied methodology describes how the

D.-J. de Grood, *TestGoal*,
DOI: 10.1007/978-3-540-78828-7_4, © Collis B.V., Leiden, The Netherlands, 2008

activities can be carried out. The best practice information described in the TestGoal approach is either a welcome addition, or can be used as second opinion. In short, an organization that is used to working with a different standard can continue to use it. If an organization is not using a test standard, TestGoal provides a proven starting point. Not applying the TestGoal approach, however, demands even more of the tester because he will have to apply the test principles even more conscientiously in order to meet the expectations of result-driven testing.

The advantages of the TestGoal approach are as follows:

- The goal is central in all test activities
- The focus on the goal and the test principles are embedded in the approach
- It's possible to start a test project quickly
- The right activities are carried out at the right time
- Best practice information and tips
- Transparent test project

In short, the TestGoal approach enables you to run a transparent and efficient test project that focuses entirely on the goal: added value for the customer. That's why we speak of result-driven testing.

4.1   Context of the Test Project

Testing is usually carried out as part of a software adaptation process.

The fact that a system has gone live does not mean that it will never change. Many a maintenance organization regularly releases a maintenance release to keep the system running and to apply small changes. Because the maintenance organization regularly produces maintenance releases, it automatically gains the experience required to analyze, specify, build and test changes. This creates a standard approach that works as long as the changes are small and the activities are straightforward.

The activities will have to be restructured as soon as the changes become more sweeping. In such cases, a project-oriented approach is often chosen. Software change projects of this kind impact the whole organization and involve several disciplines. Not only the system, but also the processes and the users are affected by the changes.

Each project has its own set-up according to the size of the project, the organization and the preferences of the stakeholders. It is, however, not unusual for test activities to be placed in a separate subproject. The below example illustrates the structure of the "Connecta" project in which testing has its own place.

Example 4.1: Connecta

The Connecta project (see also Example 1.3) consists of building a new system to replace a number of existing ones. The company has chosen this path because it expects a new system will correct many of the current errors and delays. Although the designers will leave as many existing processes as possible unchanged, a number of processes will still have to be changed. This means that users will have to get used to both a new system and new processes.

The steering committee has decided to create a number of subprojects, each with its own project manager.

- **Processes**
  This subproject defines the new business processes and ensures that they are in line with the current working method as much as possible. This subproject will also ensure that the software development project is supplied with clear information about the processes that the new system has to support.

- **Software Development**
  This subproject consists of making the modifications to the system. The system design is created according to the system requirements and the business processes. The system design forms the basis for the implementation. Module testing is also part of this subproject.

- **Testing**
  This subproject consists of testing the modifications that were made to the system. The tests are based on the system specifications and the described business processes. The tests are carried out to prevent the business processes being disrupted after the new system goes live. The tests focus on the correct working of the system and its interaction with interfacing systems. User acceptance tests are also part of this subproject. They are run to ensure the users can use the system.

- **Implementation**
  This subproject consists of introducing the system and the new procedure to the users. Because several different systems are integrated with the new system, a number of processes will work differently. It's important that the users are trained so they can continue working as they are used to.

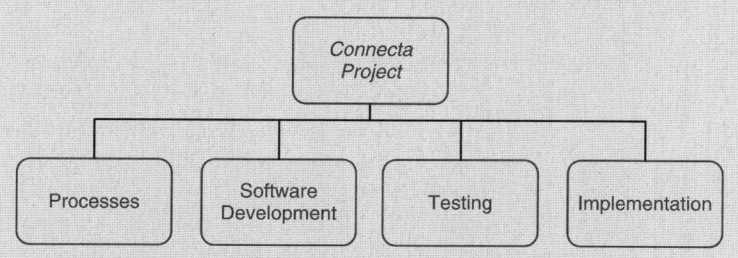

Fig. 4.1 The Connecta project is divided into a number of subprojects. Testing is a separate subproject.

As the example shows, the Testing subproject consists of several different test levels, each with its own specific system boundaries and focus. Because each test level can have a different priority, the test levels are stored separately; they distinguish themselves from each other in three areas: their aim, their focus and the system boundary.

The Aim

The aim can be different for each test level. When the aim is to demonstrate that the system works according to the system design or standards, or to demonstrate the applicability of the system, we speak of "fit for purpose." Some test levels are aimed at the chain and are supposed to demonstrate that several components successfully interact with each other, whereas the aim of other tests is to accept one component.

The Focus

The Testing subproject ensures that all of the areas of attention and risks are tested, which is why it's not enough to just test the functionality of the software. Non-functional issues such as user friendliness, performance, safety and interoperability are also important success factors. These areas of attention are called quality attributes (see the below text box). Because it is not always possible to test all of the relevant quality attributes at the same time, they are often spread across the different test levels. For example, the system test in the Connecta project is aimed at the functionality and the efficiency of data processing. The user acceptance test is aimed at user friendliness and the integration of the system with the processes. The chain test is carried out to demonstrate that the system works well with the interfaced systems.

Quality Attributes

A conceptual framework or quality model is needed to clearly express the desired quality of the software. The quality model enables a consensus to be reached about the characteristics that are important for the system. We also want to determine what the relative importance of the characteristics is. The relative importance is used to assign priorities during testing. Because it is not possible to achieve an optimal score for every quality aspect, it is not necessary to test a system for all of the aspects.

The quality requirements can be determined according to Quint. Quint is an extension of the ISO-9126 standard and describes the quality of software by means of six main characteristics (functionality, usability, efficiency, reliability, maintainability and portability) and 33 subcharacteristics [Zeist et al, 1996].

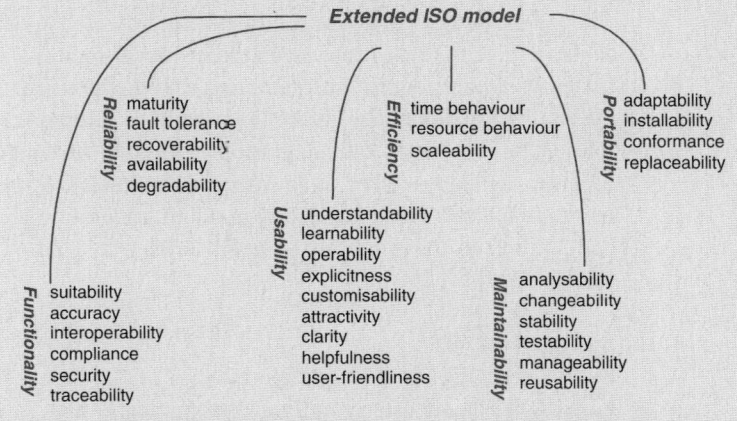

Fig. 4.2 Quality attributes according to the extended ISO-9126 Standard

The System Boundary

The system boundary has to be determined for each test level. The system boundary determines which system components are tested. The components are part of the test object. The components that are outside of the system boundary are not tested in that test level.

The test object often increases for later test levels. The first test levels focus on program components, after which the system test focuses on the complete application. If the application appears to function well, the focus shifts to how the users experience the program. In the chain test, the interaction with interfaced systems is tested.

4.2   Test Levels

As discussed in the previous chapter, the Testing subproject consists of several test levels, which distinguish themselves from each other by a different aim, focus and system size. A lot of organizations define their own test levels, as a result of which the content of the test levels is not always what the name suggests. One should be careful.

The below overview describes the most common test levels and their alternative names.

- **Module Test (MT) and Module Integration Test (MIT)**
  Module tests focus on the elementary building blocks in the code. They demonstrate that the modules meet the technical design. In modern development projects, module tests are almost always automated tests and require programming knowledge. This is why they are run by the developer before the module is released. Module integration tests focus on the integration of two or more modules. Module tests are also referred to as component tests (CT), unit tests (UT) or program tests (PT). Instead of module integration tests, the term "integration in the small" is sometimes used as well. [ISEB practitioner, 2004].

- **System Test (ST)**
  System tests usually test a whole system. The system test is a black-box test. Tests are often run based on existing system interfaces. The system test demonstrates that the system works according to the functional design. The system test is often the domain of the building party and is carried out before the system is handed over to the accepter.

- **Functional Acceptance Test (FAT)**
  The functional acceptance test is carried out by the accepter to demonstrate that the delivered system meets the required functionality. The functional acceptance test tests the functionality against the system requirements and the functional design.

- **User Acceptance Test (UAT)**
  The user acceptance test is primarily a validation test to ensure the system is "fit for purpose." The tests are based on representative scenarios from the users' daily work life. The test checks whether the users can use the system, how usable the system is and how the system integrates with the workflow and processes.

- **Production Acceptance Test (PAT)**
  The system owner uses the PAT to determine that the system is ready to go live and can go into maintenance. The system must be stable and available in the live environment. The owner also determines whether the project has supplied all of the material that is necessary to put the system into maintenance, such as system documentation, testware and installation manuals.

- **Chain Test (CT)**
  A chain test tests the interaction of the system with interfacing systems. These systems are frequently linked to each other for the fist time in the chain test. The chain test focuses on finding errors that arise when systems are not properly integrated. Because the business processes are also often checked in a chain test, the chain also tests the integration of processes with the system. The chain test is also called system integration test (SIT), end-to-end test or "integration in the large."

- **Pilot**
  The pilot is carried out prior to going live. The pilot simulates live operations in a safe environment so that the live environment is not disrupted if the pilot fails. The safe environment consists of a shadow system in a different environment in which the old situation can always be restored if problems occur. New systems usually go live with only a few users. Should problems occur, they can be solved without having too much impact on the business. Once the system has proven itself, it is rolled out across the organization and goes live.

- **Claim Testing**
  The object of a claim test is to evaluate whether a product lives up to its advertising claims.

  The manufacturer of embedded software systems or the supplier of commercial-off-the-shelf (COTS) software will try to convince his potential customers with his product. In order to distinguish his product from the competitors', he will assign several properties to his product . These properties are claimed on e. g. the products packaging or the manufacture's website.

Normally, the advertised and the implemented functionality of a software product is identical. Most software applications are truly meant to essentially do what the seller says. But in order to assure that the claims as defined by the marketing department match the true characteristics of the product, claim tests are executed.

In Example 4.1, we saw that several test levels are run in the Testing subproject. The project manager of the Testing subproject, usually the test manager, ensures that the results of the various test levels lead to the anticipated goal. Together, the results of all of the test levels bridge the gap between wish and goal. To this end, the test manager divides the areas of attention across the test levels and appoints a test coordinator to each test level. The test coordinator is responsible for the content and progress of his test project.

> **The Difference Between Test Level and Test Project**
>
> A test level is a definition of test activities with a common goal, area of attention and system boundary. A test project is a type of set up in which test activities that belong together are prepared and run. Separate test projects are frequently set up for each test level, but don't have to be. When several test levels are run by one test coordinator and the set up does not distinguish between test levels, we speak of one test project. In this case, one planning, one test plan and one test report are used.

4.3   The Details of the Test Project

Each test project has its own characteristics. The disciplines of various organizational components can differ greatly. The actual details of a test project will therefore be different for each organization. For example, compare the development department of a very technical company to the maintenance organization in the administrative sector. But all successful test projects have a number of characteristics in common[1]. These characteristics, which apply to each test project, are the basis for the TestGoal step plan. This step plan was developed to give the test project a concrete, practical shape. It assumes that there is one step plan for different environments and test levels. The actual activities can differ, but the

---

[1] "Happy families are all alike; every unhappy family is unhappy in its own way"- Leo Tolstoy in Anne Karenina

approach is the same. The step plan was written to bridge the gap between methodology and practice. It serves as a practical starting point when planning, setting up, running and concluding test activities.

4.3.1   The Step Plan

The TestGoal step plan describes which activities are carried out and which products are delivered at the end of each step. It helps the tester apply the test principles and guarantees a transparent and result-driven test project.

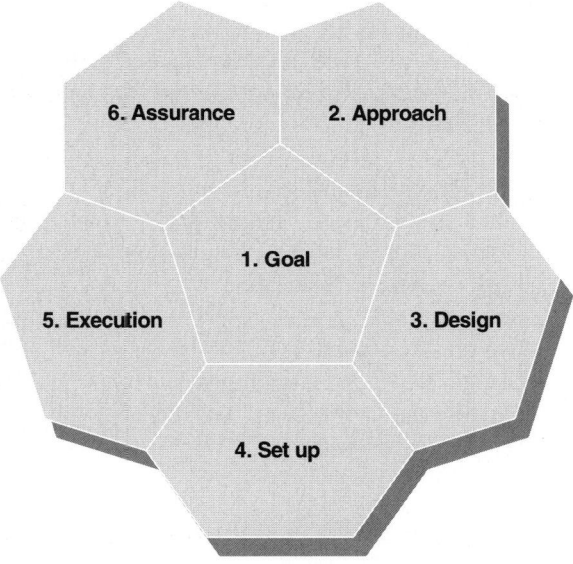

Fig. 4.3 The step plan

The step plan is displayed as a pentagon surrounded by five hexagons. The pentagon is placed in the middle and represents the Goal. The goal has to be kept in mind during all subsequent steps. The subsequent steps – Approach, Design, Set up, Performance and Assurance – are displayed as hexagons, each of which interfaces with the central pentagon. This symbolizes the contact that should exist between each subsequent step and the goal determined in the first step. The advantages of the step plan are discussed below.

- **Doing the Right Things**
  The step plan defines the activities that have to be carried out in the test project and the sequence in which they are best performed.

- **Communicating the Approach**
  The step plan provides a concrete proposal for the test approach. The step plan enables the test project to be quickly set up by indicating what the planned steps are and which products are delivered. This enables the tester to explain to the customer and the other stakeholders what he will be doing. It enables expectations to be aligned and provides a starting point for the customer to indicate what he wants.

- **Making Risks Discussable**
  The step plan is not a strait-jacket, but a reference. There's no reason why a step cannot be skipped, but there are always risks. Steps should not be skipped unless all of the risks have been discussed and understood.

- **Test Report**
  The step plan clearly indicates when which products should be delivered. Monitoring the progress enables the tester to see which products have to be finished in each step.

- **Generic Approach**
  Using a step plan makes it possible to use a generic test approach for the different test projects. This makes it easier to manage and compare the test projects.

4.3.2   *Sequence of Activities*

The step plan indicates which activities have to be carried out and puts them in a specific sequence. Going through the steps in the prescribed sequence ensures the activities are carried out smartly and efficiently. Some activities, however, can be carried out in parallel. It must, however, be clear that each step and activity has its own starting conditions. If the conditions are not met, the step or activity should not be started. For example, a test environment and a test design are needed to start running a test. The test environment can be set up earlier in the process, but it makes little sense to do so until the test design is stable. Likewise, the test design can be created earlier in the process, but it makes little sense to do so until the test environment is stable.

This is why the step plan places the execution of the test after the set up of the test environment, and the test design after the sanity check. Even if activities are carried out in parallel, the dependencies defined in the step plan will still apply.

What does a test project that was set up according to the step plan look like? To give you an idea, we'll go through the step plan for a software development project that consists of developing a new system. An example is the system test in the Connecta project. The step plan can be applied to several environments and test levels. This is why we will also indicate what changes when TestGoal is used in a maintenance environment and will look at a few other levels of testing: conformity and interoperability testing, performance testing and the testing of security.

4.4   Testing a New Program

An imaginary test project is used to describe the step plan for a new program.

Example 4.2: A test project

A test coordinator is asked to set up a test project that will be run by one or more testers. The test coordinators responsibilities consist of:

- Giving a release advice within the deadline
- Respecting the test department's rules (i. e. the organization's test strategy)
- Regularly informing the test manager about the progress, the status and the risks

In addition to his own testers (who fulfill the roles of test analyst, test engineer and test specialist), the following people are involved in the project: the project manager, the test manager, the analyst, the developer, the administrator and the user representative.

The test coordinator decides to base his project on TestGoal. He has little time and wants to be fully prepared. He can use the step plan to explain the steps and their results to everyone involved.

The scenario described in Sects. 4.4.1 to 4.4.6 indicates the steps that are taken and the people who are involved. The products that are created are printed in *italics*. The steps are described sequentially.

A detailed description is provided for each activity in Chaps. 6 to 21. The following section describes which activities are included in which step of the step plan.

4.4.1 *Step 1: Goal*

The first step in the test project is to determine the test assignment. Together with the customer, the test coordinator maps out what the anticipated goal of his test assignment is and what the business wishes are.

This step gives the test coordinator enough starting points to define the test strategy. Examples of starting points are the location of the test base, the available standards, and the names of the stakeholders, who will be informed throughout the test project and will be involved in the decision-making.

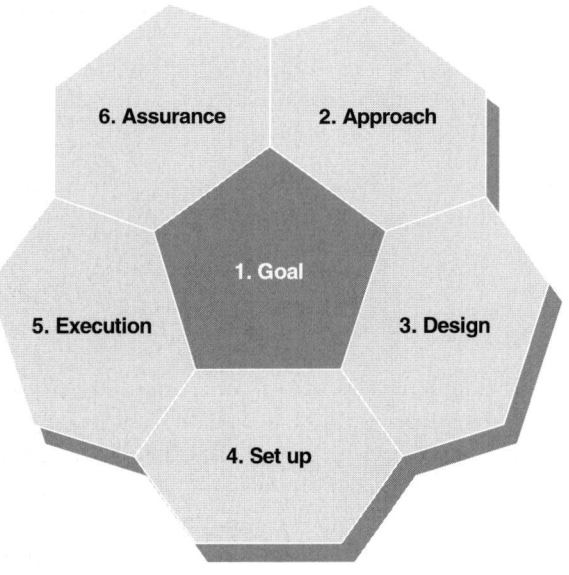

Fig. 4.4 Goal

The Goal step consists of the following activities:

1. Customer
Indicates that test activities have to be carried out. The designated test coordinator is asked to carry out at least part of the activities with his team.

2. Test Coordinator
Examines the customer's expectations, requirements and wishes and formulates them in the *goal description*. He asks for the customer's approval.

The step is concluded when the customer and the test coordinator have agreed on the anticipated goal of the test project, and if the customer trusts that the test coordinator understands the anticipated business goal.

4.4.2 Step 2: Approach

In the Approach step, the Goal described in the previous step is further elaborated. The stakeholders are asked for their vision on the product that is to be achieved. This results in a test risk analysis (TRA), which

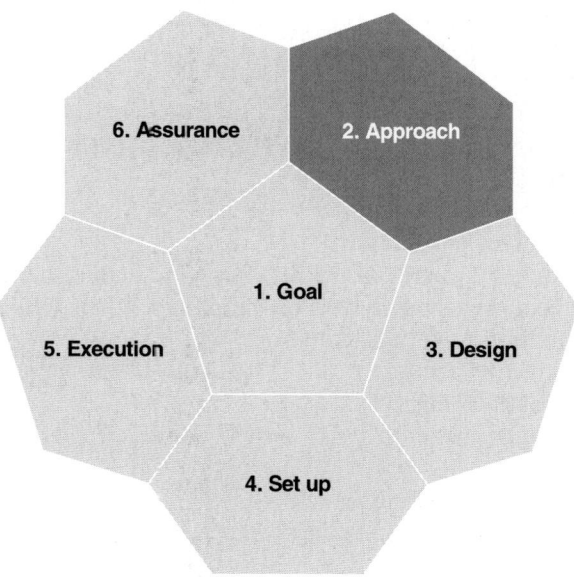

Fig. 4.5 Approach

indicates where the risks are and what impact the changes can have on the system. A budget and planning are created and the test strategy defined. The aim of the strategy is to describe how testing can effectively and efficiently contribute to achieving the goal. This is an assessment of the costs, time and risks. The strategy indicates which system components will be tested, how the tests cover the risks and areas of attention, and how the test project will be set up. The stakeholders and the customer are involved in drafting the strategy so they can indicate their interests and gain trust in the approach. The test strategy is described in the detailed test plan (DTP).

In addition to the stakeholders' input, the generic test strategy or the MTP are also used as starting points. Both documents contain guidelines for all of the test projects that are run in the project or the organization. For example, the test methodology that is to be used, the definition of the test level, the organizational structure, the stakeholders and, for example, the error-logging process. Using these guidelines as a starting point enables the test plan to be reduced to a minimum. After all, everything that is standard is already contained in the guidelines. Other input for the detailed test plan consists of the already mentioned TRA, the test budget and the test planning. Because these products must be created with diligence, they are included in the step as separate activities.

The Approach step consists of the following activities:

3. Test Coordinator
Finds out whether the organization uses a generic test strategy and familiarizes himself with it.

4. Test Coordinator
Asks his test manager about the MTP and uses it as a framework for his detailed test plan.

5. Test Coordinator
Approaches the stakeholders identified in the Goal description and asks them to share their view on the anticipated goal. He uses the *test tree* to perform a *test risk analysis*.

6. Stakeholders
Indicate that the test risk analysis correctly represents their interests.

7. Test Coordinator
Draws up the initial *budget* and *planning* and harmonizes them with the customer. If the customer agrees, the test coordinator creates a detailed test planning.

8. Test Coordinator

Uses the *goal description*, *test strategy* and *test risk analysis* to determine the approach for the test project, and describes it in his *detailed test plan*.

9. Customer

Approves the detailed test plan.

10. Test Coordinator

Kicks off the project with his testers and the stakeholders. The test coordinator explains the *test plan* and the used methodology, and highlights the most important points.

In principle, step 2 is concluded when the customer approves the detailed test plan and the stakeholders indicate they trust the test approach. The test design and the test environment can now be created.

4.4.3   Step 3: Design

The test approach was determined in the previous step. The testers can create the test design according to the detailed test plan and the test

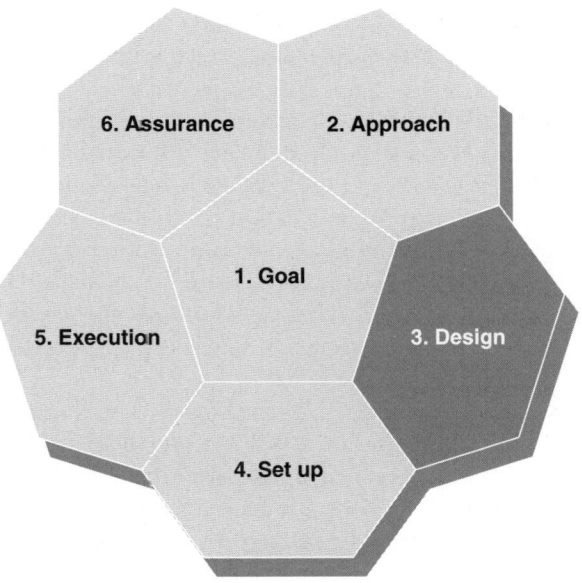

Fig. 4.6 Design

base. The purpose of the design step is to define the content of the test cases and design the test environment.

The test analysts run a sanity check to determine whether the system specifications are good enough. A thorough check is carried out to determine whether the system design sufficiently defines the anticipated goal so it can be used as a basis for test cases. This not being the case, either the system specifications are improved or the test strategy modified.

Example 4.3: A detailed test case

The test coordinator wants to create a detailed test design because the designed tests are expected to be frequently reused after the system has gone live. To ensure a good reusability, it is important that the test design has a clear structure that contains enough detail.

The sanity check reveals that the system design contains little detail. The documents cannot be used to elaborate the details of the test design as planned. The test coordinator notifies the project manager and points out the risks. He specifies that the handling of errors and exceptional situations are poorly described, which makes it difficult to determine how they should be tested. This increases the chance of surprises arising during the test and reduces the reusability of the tests. The request to improve the specifications is rejected. Elaborating the details of the specifications would delay the planning, which is not desired.

After a number of meetings, it becomes clear that the project manager will not budge. The test coordinator decides to change his test strategy. The tests are less detailed in the new strategy. To ensure that the test design still has sufficient coverage, the analysts and builders are involved. During information sessions and reviews, test cases are developed that cannot be derived from the system design. The analysts explain the design. The builders defend the implementation choices they have made.

Additional time is reserved for the test in order to determine how the tests were run. The test coordinator also expects that the testers will encounter unforeseen situations. This means that more time will be needed to fix errors. Because this will probably lead to more bug fixes, additional time is also planned for regression tests.

> The project manager is given a new version of the test plan. He's not very happy that the planning has been extended. He is, however, happy that the test coordinator is willing to cooperate to achieve the common goal. This is clearly visible, among other things, from the way he builds bridges to the stakeholders. The project manager also realizes that the builders will benefit from the information sessions – discussing their implementation choices with the analysts improves the code. The test coordinator's plan is well crafted and convincing, which is why he agrees with the new strategy and approves the test plan.

As soon as the specifications support the selected test strategy, the logical and physical test designs are created. The test design defines the tests that are necessary to determine whether the system requirements are met. The tests also check the extent to which the risks that jeopardize achieving the goal are actually realistic. Test results show where the system can be improved so the risks can be reduced, which is why the test design is written in such a way that the test results can be linked to the identified risks.

The test data is also defined. The test data contains the data that is necessary to run the tests and the data that is necessary to configure the test environment.

Although the test environment is set up and configured in the fourth step, the requirements the environment must meet are described here. The requirements concern things such as the required software, system components and possibly simulators and test tools, whether they benefit the automated test or not.

The Design step consists of the following activities. The person who carries out each activity is also indicated.

11. Test Analyst or Test Engineer
Runs a sanity check. Checklists are used to determine if work on the test design can start. The conclusion is documented in the *sanity check report*. Errors or specification issues are documented in the *review log*.

12. Test Coordinator or Test Engineer
Starts preparing the test environment. The *test environment* requirements are defined and the activities started to ensure that the test environment is ready on time.

13. Test Analyst
Uses the test strategy defined in the test plan to determine the test design techniques that are to be used.

14. Test Analyst
Applies the chosen test design techniques to the *logical test design*. Possible errors or specification issues are added to the review log.

15. Test Engineer
Takes the logical test design and develops it into a *physical test design*. The *test data* is also defined in this step.

This step is finished when all of the tests have been designed. Section 10.4.8, Assuring the quality of the test project, also demonstrates how stakeholders can be involved in the assessment of the test design so they have an overview of and insight into the tests that will be carried out, as well as into the set up of the test approach.

Experience shows that a test is frequently started as soon as the test environment and the test object are available. This can be a point in time at which not all of the tests have been defined. Executing both steps in parallel can save time, but resources can be a bottleneck.

4.4.4   Step 4: Set up

The aim of the Set up step is to prepare the test environment for testing, which can only start if the test environment is well set up. The requirements for the test environment that are established in the design step are used to purchase the required hardware and install the necessary system components. If necessary, simulators or test tools that aid the automated tests are also integrated. Once the system has been configured, the environment is tested to determine whether it is suitable for the planned tests.

If the test is automated, test scripts will be part of the available test environment. Scripts that have already been written can now be tested in the defined test environment. For an explanation on writing and implementing scripts, see Chap. 16 Test Automation.

This step is finished as soon as the environment has been provided with the version of the system that is to be tested. The smoke test report indicates whether the tests can be run. A negative conclusion following the smoke test can have various causes, for example, the system may contain one or more major errors that make further testing impossible.

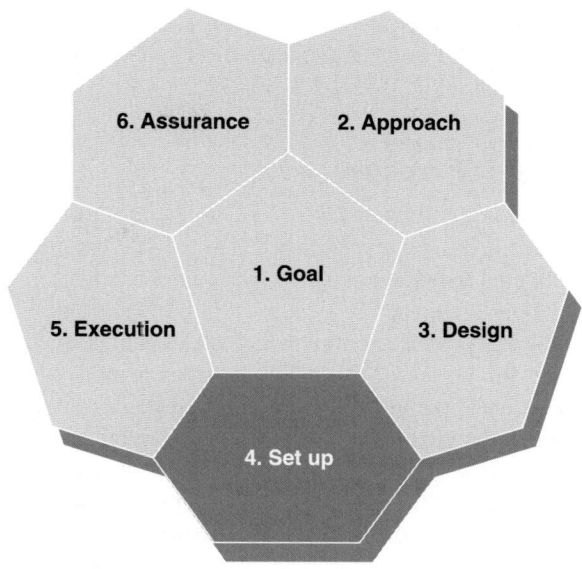

Fig. 4.7 Set up

Experience shows, however, that the cause is often due to an error in the system's configuration or installation. After the cause has been found and corrected, the smoke test is repeated.

The Set up step consists, among other things, of the below activities. The person who carries out each activity is also indicated.

16. Test Engineer or Maintenance Organization
Sets up the *test environment* so that the test scripts can be implemented and run.

17. Test Specialist (Test Automation Developer)
The test designs are used to create the *test scripts* used for automated tests. The scripts can be recorded with the automation tool's record and playback function or coded manually.

18. Test Engineer
Configures the system and performs the smoke test. Checklists are used to establish whether the test can be started. The conclusion is documented in the *smoke test report*. Errors or specification issues are documented in the Error report.

This step is finished as soon as it has been established that the test environment and the correct version of the test object are available. If automated tests are run, the tools should be integrated with the rest of the test environment and tested.

4.4.5   Step 5: Execution

Once it has been established that the test environment is available, the planned tests are run manually or automatically.

The test produces errors, which are documented. Each error is examined and the urgency with which it needs to be solved determined. A new version of the test object is made available at regular intervals. The test team reruns the tests to make sure the errors have been solved.

The stakeholders are provided with clear information about the progress of the test and the quality of the test object. The aim is to give a positive release advice on time. The release advice describes the degree to which the system contributes to the anticipated goal. A positive release advice means that the system is considered suitable for the next phase. This can be another test project or the roll out of the system in the live environment.

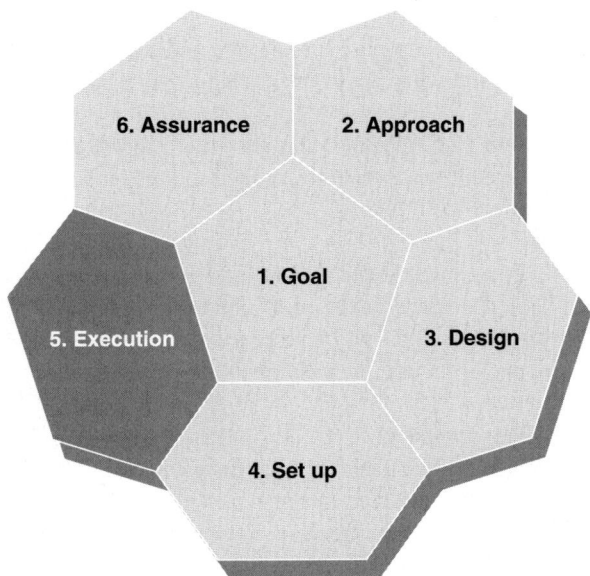

Fig. 4.8 Execution

Errors often result in changes to the system design. In such cases, the test designs need to be modified as well. For the test project, this means that the Design step has to be repeated. The tests are resynchronized with the test base. If necessary, the test data, the test scripts and the configuration are modified as well. During this activity, a check is run to establish if the anticipated goal established in step 1 is still valid. If the anticipated goal has changed in the meantime, the new insights are included in the design activities. If a new installation or system configuration is required, the Organisation step is repeated. The test is resumed after the smoke test.

The Execution step consists of the following activities. The person who carries out each activity is also indicated.

19. Test Engineer
Runs the described tests either manually or automatically. The tests are aimed at testing new functionality, or are retests or regression tests. The test engineer establishes the *test results*. *Errors* are documented in the error report.

20. Test Coordinator
Delivers the *test report* and discusses it with the stakeholders. He gives insight into the progress of the test project and into the quality of the test object. Together with the stakeholders, he establishes whether additional measures are necessary to improve the progress or the quality in order to ensure that the test project delivers the planned added value.

It is worth noting that the test report is created throughout the test project. Experience shows that test execution is commonly followed by several people. The number of stakeholders who will want to see the test report will probably be higher than during the previous steps.

21. Test Coordinator
Decides, together with the project manager, if a new version of the test object is necessary, and if yes, when. He ensures that he is informed about the errors that have been solved in the new version.

Depending on the situation, a previous step in the step plan may have to be repeated. If, however, a new version of the test object is not required, the test coordinator can draw up the final test report.

22. Test Coordinator
Draws up the *final test report* which includes, among other things, the remaining risks, recommended further activities and the release advice. It is always difficult to determine when the test run is finished.

The Execution step is finished once it has been established that the system meets the majority of the defined requirements and wishes, at which point a positive release advice is given. The test run can be stopped even if it has been decided to ignore a negative release advice and to proceed to the next phase. Chapter 18 Test Execution, discusses the question: "When are we ready?"

4.4.6   *Step 6: Assurance*

After the test has been run, the test project is evaluated in the Assurance step, and the products of the test project are handed over to the system owner or manager.

The test project is assessed during the evaluation. The result and the added value of the test project are studied and compared to the anticipated goal established in step 1. The process is then evaluated: which things went well, which measures are necessary to prevent the errors we made this time in future. The products are also assessed during the evaluation. For example, the test set is evaluated to ensure that it can be reused for a future test or regression test. Unnecessary tests are removed in order to make the regression test set more efficient. Should the

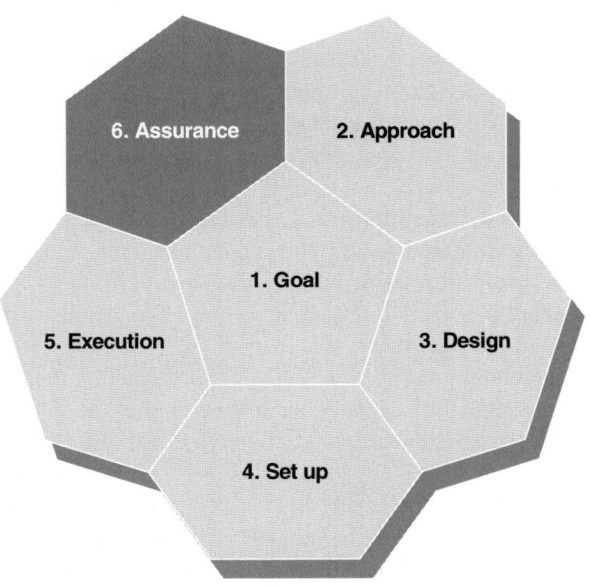

Fig. 4.9 Assurance

evaluation establish that important tests are missing, the team considers adding them.

The Assurance step consists of the following activities. The person who carries out each activity is also indicated.

23. Test Coordinator
Evaluates the test project with the test team. The stakeholders are also involved in the evaluation. All of the lessons learned are included in the *lessons learned report*.

24. Test Engineer
Determines the *regression test set* for the next testing period.

The test project is finished in the last step of the step plan. The testware is archived and is handed over to line management or the project organization so that they can reuse the products and the acquired knowledge. The handover means that all of the goals defined for the test project have been achieved. The customer discharges the test team.

25. Testers
Archive the testware and clean up the *test archive*.

26. Testers and Test Coordinator
Hand over the test archive to line management or the project organization. The test team ensures that the receiving party knows what is available and that they can use the delivered products.

27. Test Coordinator
Asks the customer to discharge the test team.

28. Customer
Discharges the test team. This completes the assignment described in the Goal description.

4.4.7   *Goal (Information and Communication)*

The focus on the goal and the test principles is embedded in each step of the TestGoal approach. It's not enough to center on the goal, it's also important to inform others about the degree to which the goal has been achieved. Throughout the test project, we provide the customer and other stakeholders with an overview of and insight into the project's progress. We inform them about the quality of the test object, about the risks that are jeopardizing the anticipated goals and about the progress of the tests. This information is provided in the *test report,* which is regularly updated and meets the stakeholders' information requirements.

A lot of test projects use progress reports. A test report is written when the test project is in its final stage. The test report is generally more formal than the progress reports. Experience shows that this is the reason why information is added to the test report that was not previously included in the progress reports. Many a customer is therefore surprised by a new context that sheds a different light on familiar data.

The test report that is used in TestGoal combines the familiar test report and the progress reports into one report, which, from the start, contains all of the information on which the final release advice is based. The final version of the test report serves as a formal test report. This prevents surprises arising at the end of the project. This approach is transparent and builds trust step by step.

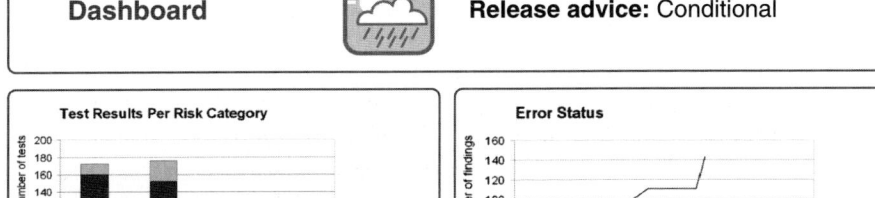

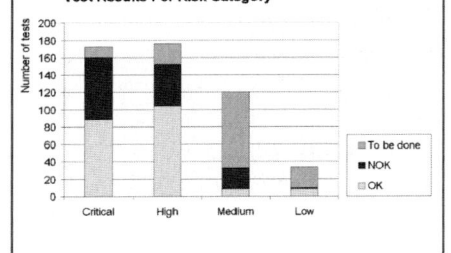

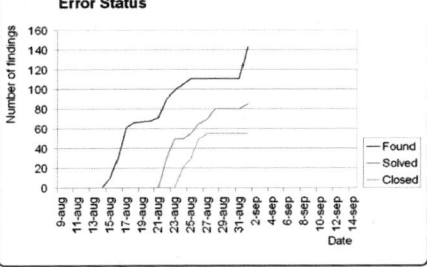

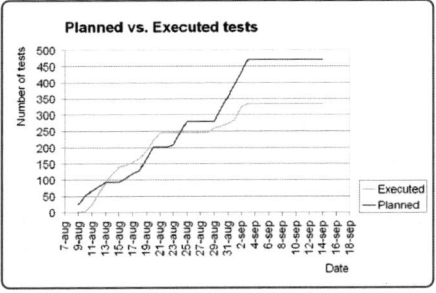

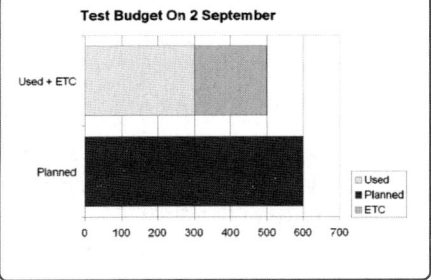

Fig. 4.10 Example of a dashboard

At the beginning of a test project, that is, during the examination of the anticipated goal, agreements are made about the test report. The content of the test report changes during the project. At the beginning, for example, there are no test results to be reported, and step-specific information is only provided during the relevant step. Chapter 20 Test Report, describes which information is relevant in which step of the test project. But not all of the information is step specific; generic information is also provided in each test report.

TestGoal recommends setting up a dashboard for the test reports. The dashboard indicates the degree to which the business requirements and wishes for the system have been tested and whether they are deemed acceptable. The status of the remaining errors and the number of tests that still have to be run provides insight into the degree to which the anticipated goal is achieved. It also shows where the remaining risks are. The dashboard provides all of the stakeholders access to the KPIs. The KPIs and the graphics that are included in the dashboard are explained in Chap. 20 Test Report.

4.4.8   Review and Acceptance

Quality assurance is an important component of any test project. TestGoal assumes that all of the products in the test project are delivered according to the organization's quality standards, which include review and acceptance, as well as a good version management system. The budget includes the time that is necessary for review and acceptance. In the test report , we specify for each product whether it is ready, reviewed or accepted. Both measures ensure that quality assurance is maintained and implemented. See Chap. 9 Test Budgeting and Planning and Chap. 20 Test Report.

4.5   Testing in a Maintenance Environment

In contrast to writing new programs in a development environment, it is not uncommon to run several short test projects when working on programs in a maintenance environment. Small changes are carried out as operational tasks that do not require a project. This applies to the changes in the software as well as to the test activities. The activities' runtime is short, but the reuse of knowledge and testware is high. The dynamics of a test project for a maintenance release are different than those of a new development project.

Example 4.4

A maintenance organization regularly produces maintenance releases, which generally contain a number of corrective changes,
but can also be used to add or change system functions. The
change requests submitted by the users and/or the maintenance
organization are revised and approved for each release. The organization has a Change Control Board, which decides, based on
the impact and urgency of the change request, which changes will
be included in the next maintenance release. Accepted changes
are implemented and tested.

The test coordinator, who works in the maintenance organization's test department, is asked to test a maintenance release. The
test is run by one or more testers. The test coordinator must:

- Deliver a well-founded release advice on time
- Adhere to the test department's rules (the organization's
  generic test strategy)
- Regularly inform the manager about the progress, status and
  risks

In addition to his own testers (test analyst, test engineer and test
specialist), the following people are involved: the manager
(release manager or change manager), the analyst, the developer
and the user representative.

The test coordinator decides to set up his project according to
TestGoal. The step plan is used as the basis for all activities. All
of the products used and created for the maintenance release are
centrally available giving everyone involved access to and insight
into the products.

The below scenario indicates where the application of a step plan in
a maintenance environment differs from the application of a step plan in
a new development project.

Step 1: Goal
The runtime for a maintenance release is often shorter than for a new
development project. The changes are generally smaller, meaning that
the Goal description has to be in place quicker. If maintenance releases
are produced on a regular basis, the expectations, requirements and
wishes placed on the system and the test process are probably already

known. If they are the same for each release, it may be possible to use
and reuse a generic Goal description.

Step 2: Approach
Because of the maintenance release's short runtime, the test approach
has to be put in place quickly. The advantage of the maintenance organization frequently producing maintenance releases is that it can use a
standard approach, which is usually described in the generic test strategy. Because much of the standard test approach can be used for specific projects, the test plan can be reduced to the exceptions that apply
to a specific maintenance release.

A test risk analysis is created for each change. A standard estimate is
not sufficient. The stakeholders and testers meet to discuss to determine
the complexity of each change request and the impact it has on the system. In this context, the test risk analysis is mainly used to determine
which system components are affected by the change. This is also referred to as an impact analysis. It is important that the test department
has knowledge of the business area the change is requested for. The test
risk analysis is used to estimate how changes have to be tested. The
thoroughness of the regression test is determined for each system component. Components that have a greater impact run a higher risk and are
tested more thoroughly.

Step 3: Design
As opposed to new development projects, there is a lot of testware that
can be reused in a maintenance situation. When making corrective
changes, errors are removed from the code, but that doesn't necessarily
mean that the test base has to be changed, which is why the tests can be
reused. The sanity check checks if this really is the case. Existing testware is checked for correctness and completeness. The risk analysis and
sanity check provide information about the parts of the testware that
have to be checked.

For the new system components, new tests are designed based on new
or adapted parts of the test base. Except for the fact that the test base for
a single change will be smaller, the sanity check does not substantially
differ from that for a new development project.

Many maintenance organizations have a standard test environment for
maintenance and bug fix releases. In such cases, the requirements for
the test environment are much easier to create because all you have to
do is check if a specific maintenance release has additional requirements. This not being the case, the test environment can be used as is.

Step 4: Set up

This step consists of preparing the test environment. At the same time, changes are made to the existing test scripts or new test scripts are added to the test set. In contrast to a new development project, the emphasis is not on defining new scripts but on reusing the already automated tests.

If a standard test environment is available, setting up the test environment is much easier, too. All you have to do is establish whether the environment is available. Any changes that are needed to test the new system components will, of course, still have to be made.

Step 5: Execution

The test consists of running the planned tests and analyzing the errors. At the end of the process, a release advice is either given for each individual change request or for the entire release.

Example 4.5: Phased testing

In the maintenance organization, testing is phased. Each maintenance release is put through a number of test levels. After the developer has changed the code, he runs a module test, after which a system test is run by the maintenance organization's test department. A module test and a system test are run for each function that is changed.

These test projects are all related to one specific change. A release advice is given for each change to indicate that the change has been correctly implemented and can be included in the maintenance release.

After the system test, a functional acceptance test (FAT) is run. This test does not focus on individual changes, but on the entire release. The FAT tests all of the functions' integration. Business processes are used to determine whether the system still supports the operational processes, such as purchasing or invoicing.

Experience shows that, despite a thorough risk analysis and system test, a change can still have an unforeseen impact. For example, making a change to a database table for one function can cause the screen that displays invoices not to work anymore. The problem analysis reveals that it was overlooked that this screen also uses the table. Or an outgoing message is rejected by the interfacing system after changing the field length of an input field. The validity check carried out by the interfacing system was overlooked.

When there is no regression, the program is released to the next test level after the FAT. The advice can be for the product acceptance test (PAT) or the chain test.

Step 6: Assurance
The evaluation of the test process has a different meaning in the maintenance organization than in a new development project. The work is less unique than in a new development project and the process is carried out regularly. This, however, does not make the evaluation less valuable. On the contrary. Because the same process is repeated at short intervals, the evaluation may be even more important than in a new development project. The evaluation can be dissociated from the individual releases, for example, by doing periodic evaluations.

It is also important that the test set be regularly evaluated. The test set is adapted based on experience gained during testing and the nature of the errors found in the live environment. The regression test set is now ready for the next maintenance release.

In the last step of the step plan, all of the testware is archived so that the test department can use the modified and additional testware for a subsequent maintenance release. In Example 4.4, the tests are run by a test department, meaning that the team does not have to be explicitly discharged because it will not be dissolved and is getting ready to test the next release.

4.6   Testing Conformity and Interoperability

4.6.1   *Introduction*

For many systems, "fit for purpose" is more important than the extent to which the system meets the system design criteria. Deviations to the specifications are acceptable as long as they support the business process. This is different for standard systems, which must meet the specified standards in order to be "fit for purpose."

Conformity and interoperability tests are run to demonstrate that the test object works according to the defined requirements and that it works with or can be connected to other systems. It is often not possible to test the test object against each target system, which is why the test checks whether the test object meets the defined standards. The standards dictate the criteria the system has to meet. There are a number of organizations that determine standards: the NNI (Nederlands Normalisatie Instituut), ISO (International Organisation for Standardization) and the field-specific IEEE (Institute of Electrical and Electronics Engineers). Governments and private companies can also define and impose standards, with or without the cooperation of other parties.

Conformity and interoperability tests are often run on systems that work in a cross-organizational infrastructure where several suppliers deliver products that have to work in the infrastructure. An example is quad-band cell phones, which have to be able to be used in different countries. Manufacturers will do everything to meet the defined standards so consumers can use their phone in any country without problems and the owner of the infrastructure can guarantee a service of constant quality.

Because embedded software systems are produced and distributed in large numbers, it is particularly important that they meet the defined standards because the products are often out of reach for the maintenance engineer, making it difficult to correct errors. This applies to cell phones as well as to bank cards. If a bank card does not work according to the standard, there is a big chance that interoperability problems will occur. Vacationers do not want to discover at their destination that they cannot use their bank card in local ATMs. Correcting an error in a bank card is a very costly activity, especially if the bank has to recall the issued cards. Banks run the risk of damaging their image, which is why it is important that such situations are prevented. It is extremely important to have an unambiguous standard, such as the EMV standard, and a related test set with good coverage.

The third situation in which conformity and interoperability play an important role, is in a Service-oriented Architecture (SOA), where generic functionality is packaged as services. The services do not have a user interface but are called by different applications. Standard messages are used to exchange services through an Enterprise Service Bus (ESB), which translates message formats, routes the messages and converts the message protocols. Translation layers are necessary because the semantics and syntax of the exchanged information can differ per system.

The idea behind SOA is that business processes are continually subject to change. A flexible IT infrastructure is necessary in order to support the business processes properly. In an SOA, generic services can quickly be combined into new applications, providing the services interact well. This is achieved by applying various open standards, such as WSDL, SOAP, XML and UDDI. In an SOA, business processes are defined by the order in which the services can be called. This is called "orchestration." To reduce the number of errors while testing the business processes, the individual services are thoroughly tested before they are integrated. The test not only focuses on functionality, but also on conformity and interoperability [van Es, et al, 2005] [Ash, 2006].

The above examples enable us to conclude that conformity and interoperability tests are often run to test functionality at the interface level and the communication protocol level.

4.6.2   *Applying the Step Plan*

The test process for conformity and interoperability tests does not differ substantially from the other test levels described in this book, which means that the TestGoal step plan can also be applied to them. There are, however, a number of differences that need to be highlighted.

Focus of the Activities
In new development projects, a relatively large amount of time is spent on devising the approach and developing the test cases. For conformity and interoperability tests, the Approach and Design steps in the Test-Goal step plan are not given much attention. The main focus is on the Set up and Execution steps. This is due to the fact that conformity and interoperability tests are often repeated for different systems. With one test set, different systems from different suppliers can be tested against the standards. This automatically shifts the focus of the activities from "devising" to "using."

In new development projects, the design and test design are often created in the same organization, stimulating communication between the disciplines. Organizations that run conformity and interoperability tests are often not involved in determining the standards. Indeed, some tests are even dictated by the standards organizations. In this case, the party running the tests limits itself to scripting and running them.

Test Automation
Conformity and interoperability tests are very suitable for automation for a number of reasons.

- **Stability of the Standards**
  Standards do not change very often. In new development projects, changing requirements have to be taken into account and a lot of time spent maintaining the test design. Because the standards are so stable, the conformity tests are easy to maintain and automate.

- **Repeatability**
  Test automation guarantees that each product is tested in exactly the same way. This adds value to a declaration of conformity and to a quality mark or certificate.

- **Stable Test Environment**
  Environments in which standards are applied generally have a stable infrastructure. This makes it possible to create a test environment that requires little to no configuration for each of the tested systems.

- **Technical Necessity**
  As previously concluded, conformity and interoperability tests are often run on interfaces and communication protocols. Their technical nature and lack of a user interface require automation in order to test them efficiently (see also Chap. 16 Test Automation).

Solving Errors
In general it can be said that merely recording findings has little added value. Value is only added if the findings are also processed. This is different for conformity and interoperability tests because they only specify where the test object deviates from the standards. Test labortories, or test labs, that offer conformity and interoperability tests as a service run the required tests. They report the results and indicate whether the standard was met, and if not, for which points it wasn't. The supplier is responsible for analyzing and solving findings. The supplier also decides if and when the modified product is submitted for retesting. Although the tester is less involved in the development of the system, this level of testing can still be result driven. Here, the added value is found in:

- Thorough and objective testing against the standards
- Clear and unambiguous reporting about the findings
- Recommending how the standards can be met

Multi-level Testing
In addition to software tests, conformity and interoperability tests also comprise tests for the hardware and the more technical aspects of a system. The standards dictate requirements at several levels alongside the software: physical, electronic, data transmission, protocol, and so forth.

All of these aspects are included in the test to ensure that a declaration of conformity is obtained.

Test Risks

The set of conformity and interoperability tests that is required for a product can depend on the product's specific settings and configuration, in which case the selection of tests is not necessarily based on the risks. When submitting a product for testing, the supplier informs the test lab of its product's specific settings and configuration. This he does, for example, in the form of an Implementation Conformance Statement (ICS). In this case, the test lab does not design its activities and their focus according to a test risk analysis, but according to the ICS.

4.6.3   Certification Tests

Certification tests are conformity or interoperability tests that are carried out with the aim of obtaining declarations of conformity, which are paired with legal and other liabilities.

When conformity or interoperability tests are successful, the product is awarded a quality mark or quality certificate. This quality mark or quality certificate is awarded by an official authority that also supervises the test. In some cases, the tests may only be run by a test lab that is accredited by an official authority. The supplier of a certified system can demonstrate that he meets the standards and their associated responsibilities. This enables the supplier to guarantee that the system works correctly with other systems. This builds trust.

> *Example 4.6: A credit card*
>
> A large bank issues credit cards to its customers. The logo on the card guarantees optimal ease of payment. The bank emphasizes that the customer can use the card anywhere in the world and does not need to worry about being able to access their money.
>
> In order to meet these expectations, every card the bank issues has to work in every ATM that bears the card's logo. To this end, a number of banks have developed a common standard for ATMs and cards. If the standard is met, the goals of the logo will have been met, and the card will work in any bank's ATM.

An independent certification authority was established to test the credit cards and their functionality in the ATMs. The authority has translated the standard to a solid test set, which was validated by the bank. A foreign branch of the bank can only issue a credit card or introduce a new ATM if the conformity or interoperability tests are successful.

The certification authority developed a standard test approach to test credit cards and their functionality in the ATMs. The approach is based on experience gained from previous test projects. The use of test tools and simulators enable the cards and the ATMs to be certified independently of each other. During the sales cycle, the certification authority explains its approach to the customer, and how the results and findings will be communicated. Because the authority's testers are familiar with the test approach, the preparation time is minimal. As a result, testing can start as soon as the test object has been delivered.

For some critical systems, the official certification tests have to be successful before the system can go live. A supplier of such systems has to have a certificate, or several sub-certificates, in order to market the product. Sometimes a certificate is only granted if the supplier has obtained all of the required sub-certificates by running, or contracting someone else to run, the tests at various test labs.

For important certificates, the way in which the tests are run is subject to requirements that are imposed by, for example, the government or an umbrella agency. Certification tests are therefore run by specialized test labs. Before a test lab is allowed to run official certification tests, it has to be accredited. In order to obtain the accreditation, the test lab first has to demonstrate, by means of an audit, for example, that the requirements of the certifying authority have been met. In order to keep the accreditation, the test lab is periodically reassessed. In the Netherlands, the Dutch Accreditation Council controls the test authorities. At the international level, it has been agreed that each country have such an authority, which also has to meet the specified requirements.

Less important certificates have less strict rules. Organizations can define their own set of requirements, like the banks in Example 4.6. In this example, the cooperating banks are the authority that draws up the requirements the test lab has to meet in order to be accredited. In some cases, the product supplier runs the tests and submits the test results and required reports to the certifying authority for approval. The certifying authority examines and checks the test report . If all of the requirements are satisfied, the authority grants approval in the form of, for example, a quality mark or a quality certificate. In some cases, the product supplier can grant itself the certificate after running the required tests. By doing this, however, the supplier accepts legal liability if it can be proven that the tests were not run correctly.

As the above shows, certification tests can lead to official declarations of conformity in a number of ways. The way in which conformity is confirmed depends on the requirements that apply to the product and the involved organizations. It is important that recognized test labs run the tests and that the labs are controlled according to established standards. This is the only way to guarantee the trustworthiness and usability of a declaration of conformity.

4.7   Testing Performance

4.7.1   *Introduction*

An important aspect of the quality of an IT system is its efficiency. System efficiency is defined as the degree to which the system is able to deliver the right performance, depending on the number of resources used under the specified conditions [ISTQB, 2005]. A common description of the performance question is: "If a lot of users are using the system, will it be fast enough?" Performance tests are used to determine the system's efficiency and to answer the performance question. Performance tests are specially developed to test efficiency and are run to demonstrate that the system responds well to long and intensive use.

Below is a popular definition of performance testing:

> Performance testing is the validation of requirements with respect to time and speed and the use of a specific information system's resources [Pol et al 1999] [Siteur 2005].

Performance tests are very important, especially when the system is expected to have a large number of users. This is demonstrated by, among other things, a study that was conducted by Akamai and Jupiter Research in 2006. The study concluded that the average visitor of a Web store will wait no more than four seconds for a Web page to load. If the page takes longer to load, the user leaves. The study also found that one-third of the Internet users who have a bad experience with a site never return. Three-quarters of the visitors indicated that they would probably not buy anything from such a site [Akamai 2006]. A slow Web site can damage the company's reputation and result in significantly lower sales. That's why it's very important to make sure that the system has a high enough efficiency. The aim of performance testing is to provide timely insight into a lack of performance and to produce reliable results.

This not only applies to Web-based systems. In general, it can be said that the users' acceptance of a system depends partly on its performance. If users find that the system they need to work with is slow, they will either completely ignore it or declare it as not usable. Systems that do not reach the expected performance level are generally not usable.

With a view toward the anticipated business goal, performance is taken into account in the design, the development and the testing. This starts by defining the performance requirements. The results of the performance tests are used to make a statement about the degree to which the requirements have been met and the likelihood that the system will contribute to the goal.

4.7.2   Applying the Step Plan

The performance test process is not substantially different than the other test levels described in this book. The TestGoal step plan can therefore be applied, under consideration of the below.

Performance tests can be run during the entire software development project. In practice, however, performance tests are mainly run later in the project as part of the acceptance test because a representative test environment is needed. The live environment's configuration and hardware are often unknown until late in the project. The live environment cannot be simulated until its configuration and hardware have been determined. Another reason is that the performance is best run on a functionally correct system, which is why performance tests are often run only after the system test or FAT. Although these are valid arguments, it is wise to run the performance tests as early as possible. The first performance tests could be run during the integration of various components, that is, during the integration test. Performance problems can have deep architectural causes [Anderson 1999]. The later these causes are detected, the harder and the more expensive it will be correct them [Boehm 1981].

Experience shows that the performance requirements are often not available or are incomplete. Performance tests frequently turn into performance measurements and end up as measurement results rather than in a release advice. The measurement results need to be analyzed after the test has been run. The stakeholders can use the outcome of the analysis to say how they rate the measured performance.

Improvement actions may have to be performed on the test object. Such improvements are often implemented by the multidisciplinary team that was involved in the measurements and the optimization of the performance. The performance tester works with the team's analyst, IT architect and builder. After the improvements have been implemented, the team provides insight into the effect of the improvement actions by repeating the tests.

The above shows that it is difficult to specify the quality of the test object in the test report . This is why an additional report is often created, which indicates how the measurements were made, what the measurement results are and how the performance can be improved.

The results of the performance tests are used to improve the system, but they also have another use. The quantitative character of the test results is suitable for service level agreements (SLA). The results give an indication of the performance, which can be used as a limit value in the SLA.

4.7.3   Test Design Techniques

Performance testing also tests the system's response times. Different terms and test notions are often used to describe the same thing. To ensure clarity, the following notion is used in TestGoal.

> The term performance testing is a general notion or collective term for the testing of response times [Siteur 2005].

In performance testing, the following test design techniques are distinguished:

Load Testing
Determines the optimal processing time at a representative system load.

Stress Testing
Determines the maximum load the system can bear and provides insight into the dependency between the load and the processing time.

Reliability Testing
Demonstrates the system's reliability by letting it run under a representative load for a longer period of time.

Concurrency Testing
Demonstrates the system's reliability by establishing that the various processes do not have a negative influence on each other.

Each test design technique has its own focus, and each test design techniques enables different bottlenecks to be found. The different test design techniques enable performance testers to look at the system in different ways. The last two test design techniques focus on the "reliability" aspect, but they can reveal problems that can have consequences for the response times and hence also say something about the efficiency. This is why these techniques have been categorized as performance tests. In addition, these test design techniques use the same scripts and tools. A tester who is running load and stress tests can often run the reliability and concurrency tests with little additional effort. This is an efficient way of obtaining information about how the system works.

4.7.4   *Test Tools*

Test tools are indispensable in performance testing. Performance testing consists of running a large number of transactions over a longer period of time, making it impracticable to run the tests manually.

Performance test tools monitor the test object under various usage conditions and report on its behavior. The tools are often based on automated tests [ISTQB, 2005].

Tools are used to create a heavy load. The load is determined on the basis of the performance requirements or the expected use. Performance tools enable profiles to be created based on parameters. Examples of parameters are the number of simulated users, the number of repetitions in a test, and a specific period within which the tests are run in cycles.

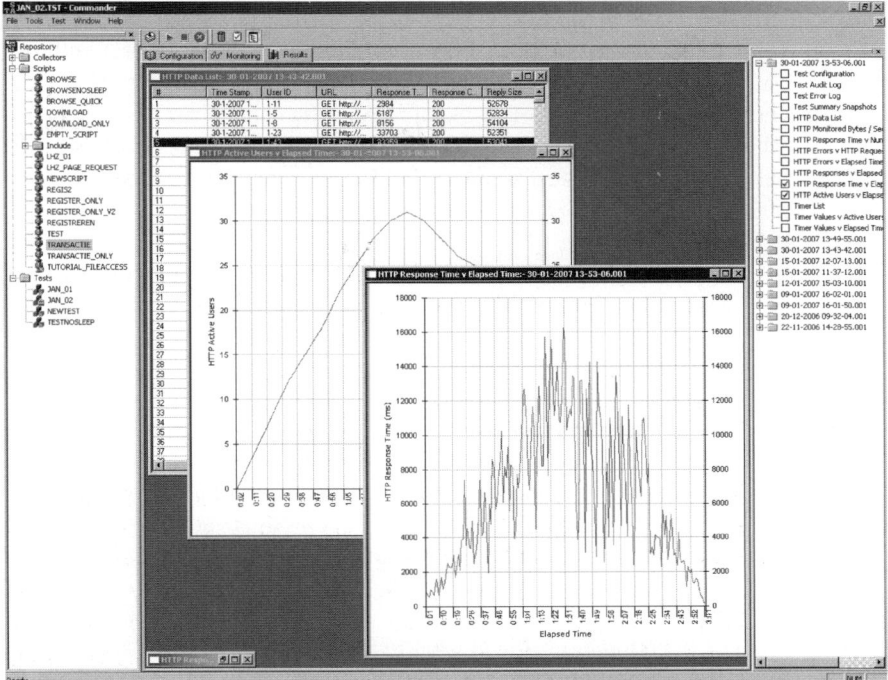

Fig. 4.11 Results of a performance test run on a Web application using the open source tool System Testing Architecture (OpenSTA). The main screen displays the measurement data, the load imposed and the measured transaction time. It is clearly visible that the transaction times increase with the number of users using the system.

Tools are also used to monitor the usage of system resources, such as the processor. Monitoring the used system resources enables us to determine the breaking points or bottlenecks. In addition, performance tools generally produce a lot of measurement data, which is stored in a database or written to log files. Performance test tools can analyze and graphically display measurement data.

Performance tests are characterized by a lot of measuring and calculating. Due to the strong technical nature of the tests, we see performance testing as a separate specialism.

4.7.5 Dependencies

When assessing the results of the performance test, it is important to know to which extent the results are representative for the live environment. In order to obtain representative results, it is important to run the test in an environment that closely resembles or is identical to the live environment. In many organizations, the acceptance test environment is used because it simulates the live environment best. There are, however, a number of other factors that influence the similarity between the two environments that a performance tester will have to be aware of and include in the test design, the test environment, the test itself and the test report. Below are examples of such factors:

- **Test Configuration**
  The test configuration must be as close to that of the live environment as possible. If it differs, the impact of the difference will have to be included in the test results.

- **Operational Load Profile**
  A load profile will have to be created based on the users using the test object and the transactions they will be running. Determining the frequency of use and of the transactions will ensure the system load is simulated as accurately as possible.

- **Batch Transactions**
  The operational profile is used to simulate transactions that occur in a test session at a specific point in time. Batch transactions enable the transactions to be simulated as accurately as possible.

- **Non-intrusion**
  When the test tool and the test object are in the same configuration, the test tool will influence the performance of the test object. To eliminate these effects, a test should be created as non-intrusively as possible.

- **Test Data**
  There must be enough test data so that advanced queries can be run and considered in the performance results.

- **Database Size**
  In addition to sufficient test data, the database size will have to be included in the performance test. Queries run faster on smaller databases than they do on larger databases. It must also be taken into account that live databases can grow quickly.

- **Indirect Influences Additional Software/Hardware**
  The test environment will not contain any other systems that influence the test configuration. When presenting or assessing the performance, one must take into account that users will probably be using other software alongside the test object, which can influence the performance or that the hardware will also be used for other ends.

When one or more factors need to deviate, the deviations will also have to be considered in the test results. Factors such as error analysis and significance will also play a role (see Sect. 12.6.8 Load testing). All of the factors will have to be mapped before a reliable statement can be made about the performance.

4.8   Testing Security

4.8.1   *Introduction*

One of the most important technical challenges nowadays is to produce secure software as security is the key limiting factor in deploying information technology. Security testing can help to build a secure application.

It is quite hard to give a definition of security testing that covers this term completely, as security testing encompasses a wide area. A general definition would state something like: the process of determining to what extent a system as a whole secures confidentiality, integrity and availability of information. Here, a system refers not just to an IT system, but also to procedural and organizational components. Confidentiality is the prevention of intentional or unintentional unauthorized disclosure of contents. Integrity concerns the guarantee that the message sent is the message received and that the message is not intentionally or unintentionally altered. The concept of availability ensures that information is accessible. [Hansche, 2003]

The primary goal would be to achieve a satisfactory level of confidence. This way testing increases trust in the system and helps to prevent damages caused by an insufficient security level. It makes the current system more secure by mitigating current risks and it assists in future improvements of the system's security.

4.8.2   *Approach*

Within security testing, both hacking and testing principles come together. The security tester therefore should have knowledge of both testing and security, a strong focus on detail, and a wide knowledge of current trends in security.

In order to find security errors, the security tester should try to adopt the mindset of a hacker who is trying to break the application. The hacker is assumed to have access to all system components that are not protected by authentication rules. Creative thinking can help to determine which set of data may cause a system to behave in an insecure way, or assumptions made by developers that can be exploited [OWASP].

As well as being creative, a security tester should have a broad knowledge of the best practices and known threats. He should also have knowledge of the organization's security policy. Even scenarios of low probability should be tested.

Security Testing Terminology

Threat: An environment or situation that could lead to a potential breach of security. Examples of threats: Phishing, Worms, Viruses, Power loss and Fire.

Vulnerability: An existence of a software flaw, logic design, or implementation error that can lead to an unexpected and undesirable event executing bad or damaging instructions to the system. Examples of vulnerabilities: Using insecure coding techniques and Insufficient input validation.

Exploit: A defined way to breach the security of an IT system through a vulnerability. Examples of exploits: Cross site scripting (XSS), SQL injection and Buffer overflow.

Attack: Occurs when a system is compromised based on a vulnerability.

Security analysis: Looking for and prioritize threats.

Target of evaluation: System, program, or network that is the subject of a security analysis or attack. The test object.

When it comes to testing, it's all about the details. This especially goes for security testing as the smallest leak often leads to the biggest security threat. A minor exploit which individually does not represent much risk, may lead to a security breach when aggregated. In security testing, the structured method of software testing should be followed. By remaining well-organized and persistent, the executed tests will give information about risks, weaknesses, information leaks and vulnerabilities. The challenge for a security tester is to translate this information into clear test reporting. Stakeholders are given insight into the domains where not only the anticipated goal is endangered but also into the domains where the systems security can be trusted.

4.8.3 *Applying the Step Plan*

Taking the previous remarks into account, the essentials of the test process, as described in the TestGoal step plan, can be adopted for security testing as well.

In step 1, the Goal description is being formulated and approved by the customer. Since gaining access to the company's operational systems is illegal in most countries, it is important to have explicit authorization. This authorization can be included in the goal description, or otherwise in a separate contract.

Security testing is like any other test level; it is a risk-based activity. Step 2 of the step plan describes 1D and 2D TRA. It is the 2D TRA that is especially useful for determining the focal points for the security tests, as this is likely to detect non-functional risks at a later stage that are not addressed in the test base.

The test design techniques used in the third step of the step plan (design) are also applicable for security testing. However, Security tests don't usually take into account the "normal" way of using the system. For example, the security tester will test a web server using the client (like a browser), but will also try to gain access through the back door, just as a malicious user will. Section 12.7 describes how the traditional test design techniques can serve this purpose.

While executing the tests, a lot of creativity is required from the tester. Since our aim is to provide structured information on the security threats, a physical test design is used. However, the test design will allow the tester some degrees of freedom, in order to explore the systems weaknesses.

During test execution (step 5), test reporting is done like in any other test project. One should realize that security testing is an ongoing process, as the environment in which the tests are performed is continuously changing. Hackers keep finding new ways to exploit weaknesses. It is the responsibility of the security to make the customer aware that a security test represents a snapshot of the system: it provides a profile of the system concerning known vulnerabilities, known weaknesses and known configurations at that time [Herzog]. The security level of the system alters as soon as changes are made to the system or its configuration. Therefore the used configuration is included in the test report. It should list the configuration of the system, the versions of test tools being used and, of course, the date when the security test was performed.

The ISO 9126 standard contains several quality attributes that encompass security related items (e. g. performance testing, maintainability, etc). When security is a main asset of the business, the ISO model is insufficient. In this case security is an attribute for itself.

CISSP, which stands for Certified Information Systems Security Professional, is a leading certification for security professionals. Within CISSP ten domains are defined that indicate how broad the field of security testing really is. All these domains together are called the Common Body of Knowledge (CBK). The CBK can act as a knowledge base for the security tester. The ten domains are:

- Information Security and Risk Management
- Access Control
- Telecommunications and Network Security
- Cryptography
- Security Architecture and Design
- Operations Security
- Application Security
- Business Continuity Planning and Disaster Recovery Planning
- Legal, Regulations
- Compliance and Investigations and Physical security

For each domain security tests can be set up.

Testing security should be organized as a separate test project, thus ensuring that the security tests take place at all test levels.

For large projects, security is often organized as a separate test project. The Test coordinator has the responsibility of ensuring that all security related test activities are aligned. To do so, he will assume the responsibility of embedding the security tests in the master test plan. He and his team will take care of the preparation and execution of the tests, as well as making recommendations for improvements. These later often show to be indispensable as the security tester is often the only person skilled enough to analyze and provide solutions for security exploits at the client's site.

Usage of automated and specialized tools are necessary when performing a security test, as testing against known vulnerabilities by hand would be highly inefficient.

As for the domains, a suggested order in testing security is: first review compliance to legislation and the organization's regulation and policy; second, check configurations of security measures (i. e. firewall); third, use automated tools to uncover known vulnerabilities; and finally, combine creativeness, best practices and knowledge of known threats to produce and execute a set of test cases.

5 Getting Started

The previous chapters briefly described the whole test process for the different environments and different test levels. A tester's mindset and expertise were also discussed. The following chapters describe the test process step by step and look at each activity in the TestGoal approach in detail.

Chapters 6 to 21 discuss the six steps in the step plan. Each step's aim, activities and deliverables are explained on a separate page between the chapters.

To make this practical part of the book accessible to as broad an audience as possible, each activity is described in a separate chapter. The activities are described in the same order in which they are carried out in the step plan. This makes it a useful quick reference guide for testers.

D.-J. de Grood, *TestGoal*,
DOI: 10.1007/978-3-540-78828-7_5, © Collis B.V., Leiden, The Netherlands, 2008

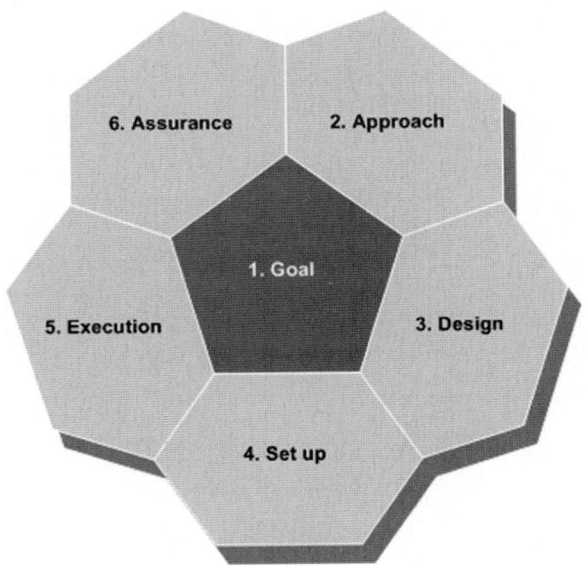

The first step, Goal, is central to the step plan. In this step, the antici-pated goal is assessed. The expected results of the test project are estab-lished and the business goal defined. Understanding the anticipated business goal helps the test coordinator organize his test strategy in such a way that the system's real success factors are tested.

The Goal step consists of the following activities and products:

Activity	Product
Goal assessment	Goal description

6   Assessing the Anticipated Goal

6.1   Introduction

Before a test project starts, the stakeholders usually have an idea of the test activities and specific expectations concerning the result. When the project starts, it's important for the test manager or test coordinator to know what the actual test assignment consists of and what the expectations are. The first activity in the step plan focuses on mapping out the expectations and assessing the anticipated goal. This activity's deliverable is the Goal description.

The Goal description is sometimes compared to the project brief (the project proposal) used in the PRINCE2 method [Hedeman, 2000]. The project brief describes in summary what the reason for the project is. It also describes, among other things, the business case and the quality expectations. The project brief is written by the customer and the project manager, and defines the scope of the project to be started. A known risk in the use of the project brief is that the project manager is not always involved in its creation. Consequently, he does not always have insight into the project's expectations. This is one of the reasons why the project manager often uses the project brief to reduce his responsibilities and to keep problems outside of the project. In practice, it is not uncommon that project managers protect themselves by hiding behind the project brief [ZBC], [Steenberg, 2005].

The assessment of the anticipated goal is a relatively small, but extremely important step that forms the core of TestGoal. In contrast to the projects themselves, this step focuses not only on the formulation of the assignment, but especially on understanding the goal the organization wants to achieve. The Goal description establishes the agreements

D.-J. de Grood, *TestGoal*,
DOI: 10.1007/978-3-540-78828-7_6, © Collis B.V., Leiden, The Netherlands, 2008

between the customer and the test coordinator. Clear agreements are important, but the Goal description should not be used to protect the test coordinator. The Goal description is defined to gain a clear understanding of the real reason for the project. What are the business goals and what are the anticipated effects the project will have? It is important that we as testers understand how we can contribute to these results with the testing activities.

6.2   Aim of the Assessment

At the beginning of a new test project, the first activity is the assessment of the goal to be achieved. A description of the assignment may already exist, for example, as part of the proposal or the customer's request for resources. If a description is not available or leaves doubts about the anticipated goal, the test coordinator will have to create one himself. In either case, the test coordinator has to communicate with his customer and get a picture of the following points:

The Anticipated Goal
As we saw in Sect. 1.2, the business goals are translated into system specifications in a software development project. A common assumption is that if the system meets the specifications, the business goals will have been achieved. The system requirements will quickly take a central place in the assessment of the anticipated goal. After all, the test assignment will often be summarized as: "Demonstrate that the system meets the system requirements." The power of TestGoal is that it includes two other goals: the anticipated goal of the business and of the test project itself.

In many cases, more is needed to achieve the business goals than the modification or creation of an IT system. The business processes, too, will have to change, and the users will have to be trained – they are the people who will have to work with the adapted business processes and the adapted system. It's important to realize that the goal of the software development project is only part of the total operation.

A test project also has an anticipated goal of its own. The test project will have to provide insight into the quality of the system. In addition, the test project will also be evaluated based on the way in which this insight is provided and the deliverables the project yields. These goals also have to be taken into account.

The Reason and Project History

In addition to the anticipated goal, the reason for the project and the project history can provide useful information. Think of issues and politics that may have played a role; they also often provide information about problems the organization is trying to solve, as well as about common pitfalls.

Example 6.1: Project history

During the assessment of the anticipated goal, the test coordinator questions the customer about the reason for and the history of the project. The customer explains that similar initiatives had been started in the past, before the company was reorganized. One of the reasons why the last attempt had failed was because the users had not been involved in the project. A system had been built, but it didn't tie into the workflow and was never completed. He explains that for this project, the users were consulted and their opinions included in the current project approach. A user meeting will be held once a month to discuss the most important system issues. The test coordinator can use this user platform to involve users in the test project.

The Assignment

While discussing the assignment, the customer describes what it consists of. It is also wise to consider the customer's expectations. The success factors are also established insofar they have not already been embedded in the anticipated goal. The success factors formulate the points at which the test project is evaluated. In principle, they are the customer's answer to the question "When do you think the test project is a success?" The test project is also delineated: which system or system component will be tested? Has the test level been determined or does the entire test strategy still have to be established? When security tests need to be executed, explicitly ask for the customer's permission to perform these tests. Gaining unauthorized access to the live systems of a company is illegal in most countries, even if you are a security tester.

The Expected Timeline and the Available Budget for the Project and Test Assignment

For many test projects, a deadline and a budget will have been determined before the project starts, for example, during the creation of the project plan. Experience shows that this happens in a lot of projects without consulting the test expert. Regardless whether the tester believes the budgeted time and money is realistic, the customer believes that it *is*. This budget is the starting point for our strategy and planning.

Reporting

During this step, agreements can be made about the progress report. The test plan will define how progress is reported. In complex projects, it can take a couple of weeks before the test plan has been completed. In such cases, it is wise to agree on the way the progress will be reported until the test plan has been completed.

After discussing the above-mentioned topics, the test coordinator creates the Goal description; this is the formulation of the assignment that is used as the starting point for the further set up of the test project. Note that the Goal description is a generic description that is no longer than one or two pages. The next step of the step plan consists of defining the test approach that is needed to achieve the anticipated goal. During the creation of the test approach, certain assumptions may turn out to be unrealistic. If, for example, it turns out that the initial timeline is not feasible, the test coordinator will look for a solution together with the customer.

6.3 Goal Description

The Goal description contains the following elements:

Project Name
The name of the project.

Anticipated Goal
A description of the goal that is to be achieved. Include what the business goal is behind the decision to create a new application or adapt an existing one. Also include what the expected added value of the test project is.

Customer
Name of the person responsible for the test project.

Accepter of the Assignment
Name of the test coordinator or of the test department.

Scope
Description of the system to be tested (test object or SUT). Indication of the system boundaries.

Test Level
The test level that is applied to the system, for example, module test, module integration test, system test, functional acceptance test, etc.

Expected End Date/Available Budget
Date on which the test project should be finished and the budget that is available.

Task Description
Short description of the test assignment and the success factors.

Agreed Reporting
Agreement on how and at which frequency the results of the test project are reported to the customer and the stakeholders. The report should also contain the elements that are included in the test report (See Chap. 20 Test reporting).

In the Connecta project (see Example 1.3), a new system is developed that will replace a number of existing systems. The below text box displays the Goal description that the test coordinator wrote together with his test manager.

Example 6.2: Goal description Connecta system test

Project Name
Conneta system test

Anticipated Goal
The new system will replace the five systems that are currently in use. The functionality of these five systems will be integrated in the Connecta module. The integration is expected to increase the efficiency of data processing and reduce errors. The current operational activities will be able to be carried out with fewer FTEs. In short:

- The system will process data more efficiently
- The number of errors that have to be fixed manually will be reduced
- Fewer FTEs will be needed to carry out the current operational activities
- The project will ensure that new system works well with interfaced systems. The project will also ensure that the existing

workflows are adapted to the new situation and that the users are trained.

- The test project makes a statement about the quality of the system and the risks at the launch. The goal is to ensure that business processes are not severely affected after the new system has gone live.

Customer
Connecta test manager (Yasmin Hassouni).

Accepter
System test test coordinator (David Bloom).

Scope
Connecta module.

Test Level
System test.

Expected End Date/Available Budget
The chain test for the first increment is planned to take place before September 2007. The complete Connecta module has to go live before November 1. There is enough budget for two testers and 1 FTE test coordinator until December 1.

Description of the Assignment
Organize and run the system test. Demonstrate that the system meets the system specifications and can replace the five existing systems. The goal of the test project is to give a release advice for the chain test, which will test the interaction between the systems. The test project should also provide information as to whether the new system will actually reduce the number of FTEs. David Bloom needs to stay in touch with the UAT test coordinators (in order to guarantee that the systems support the processes) and the CT (next test level).
The system test:

- Indicates whether the system is of good enough quality to start the chain test
- Indicates the extent to which it is realistic to expect that fewer FTEs will be required
- Provides a documented test design that can be reused for future regression tests
- Is run according to the guidelines in the generic test strategy

Agreed Reporting
The meetings and reports will be defined in the DTP that David
will create. The test plan is expected to be ready in two weeks. It
has been agreed that until that time, David will e-mail a report to
Yasmin every Wednesday and Friday, which will include:

- The names of the people he has talked to
- The status of the products (WBS, planning, TRA and test plan)
- The bottlenecks and risks

The goal assessment enables the tester to familiarize himself with the
activities that we know will occur in the step plan. By gathering information now, he will be able to quickly start with the test risk analysis,
the test budget and the test plan. The information he does not include in
the Goal description will be elaborated in the preparation phase of the
test project. Together with the goal description, this information is repeated in the test plan. The test plan is described in Chap. 10.

6.4   Information Gathering

The test coordinator starts defining the test approach as soon as the
Goal description is ready. The time needed to define the test approach
can be reduced if all of the relevant information is available. This
is why it's useful to obtain the necessary information during the
assessment:

6.4.1   Product Development

Test Object
Which system has to be tested? What are the system boundaries? What
should the system do, what are the risks and concerns?

Tip: Use a graphic to indicate the system boundaries and their
relationships. For an example, see Fig. 10.01 Context diagram in
Chap. 10 Test Plan.

Test levels

Which tests have to be run? For example:

- The test levels in the V-model: module test, module integration test, system test, functional acceptance test, etc. (see Sect. 4.2 for an overview of test levels).
- The quality attributes (in Quint): will performance, usability and security be tested as well as functionality?
- If the assignment is for one specific test level, how are the other test levels handled? Who is responsible? Are all of the known risks covered?

Relationship with Other Systems

Are there external dependencies that can influence the progress or the success of the test project, such as interfaces to other systems?

Development

How will the system be developed? Will specific techniques or development methods be used? Will the program be developed in-house, or will it be outsourced?

Maintenance

How will the handover to the maintenance organization take place? To whom will the products of the test project be delivered and how will they be used after the project?

Are there any relevant products the maintenance organization can make available for the test project? For example, a regression test set from an earlier project.

6.4.2   *People*

Owner of the Anticipated Goal

Who has the (biggest) interest in the anticipated goal?

Customer

If the customer is not the same person as the owner of the anticipated goal, who is the customer? Are there any other parties who can be considered as customers, or as people who want to influence the goal, such as external contractors, customers or users?

Stakeholders

In addition to the owner of the anticipated goal, more parties may have an interest in the software development project or in the test project. Who are they and what interests do they have? How can they be contacted?

Experts

Who are the experts, and in which fields? For example, the analyst who can provide information about the design, or the senior employee who has a lot of subject-matter knowledge. Can these experts be consulted during the project?

Test Team

Has a test team been put in place? What are the team's limitations? What quality/knowledge/training does the test team have? Are there enough resources to run the tests?

6.4.3 *Guidelines and Documentation*

Guidelines

Are there any project or organizational guidelines the test project must adhere to? For example, a test methodology that is used: a master test plan or requirements that are imposed by the umbrella QA policy. Also check the organization's security policies, and possible liability.

Tools

In addition to compulsory guidelines, an organization may have procedures, standards or tools that can save a lot of time because they save reinventing the wheel, for example, in the area of configuration, change and error management. Does the test team have experience with these tools and processes?

Documentation

What documentation is available and relevant? Have system acceptance criteria been defined? Has the test base been defined? Is there a project plan that defines everything that relates to the test assignment?

Terminology

Which terminology is used? Many projects or organizations have their own definitions for seemingly familiar terms. Bear in mind that the definition of a term may not be the one you're familiar with.

With the answers to the above questions in mind, the test coordinator should be able to create an initial picture of the test assignment. The test coordinator also knows who he can approach for more detailed information. The Goal description and the additional information create the basis for step 2.

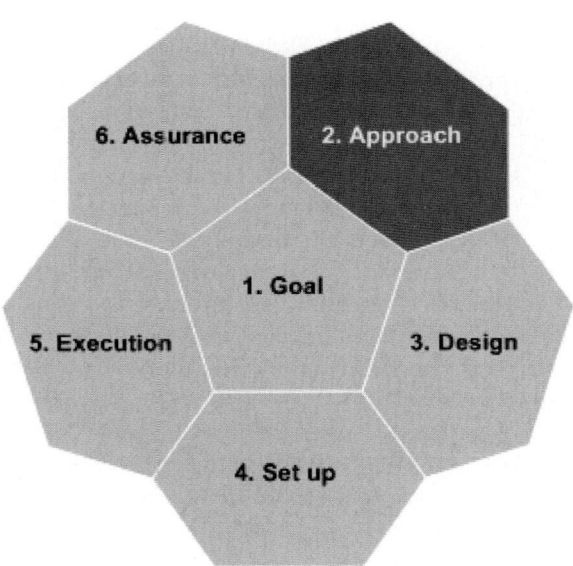

The Approach step is based on the anticipated goal that was established in the first step. The TRA and the available time are used to translate the anticipated goal into a concrete test strategy. The product of this step is a detailed test plan that includes the TRA, the planning and the test approach.

The Approach step consists of the following activities and products:

Activity	Product
Execute TRA	Test tree Test risk analysis
Budget test assignment	Budget Generic planning Detailed planning
Determine test strategy	Test strategy (in test plan)
Create test plan	Test plan

Step 2 Approach

In step 1, the anticipated goal was assessed and the test coordinator has obtained the information that will help him create a good approach for the test project. The test strategy describes how the test coordinator will effectively establish whether the anticipated goal has been achieved. In order to create the right test strategy, the following components are combined into a consistent whole in the test plan:

- **Goal Description: The Goal**
  The Goal description is the formulation of the assignment in which the test project's anticipated goal is defined.

- **The Test Risk Analysis: Threats to the Goal**
  The TRA identifies the areas of attention for the test object and estimates the risks for each of them. Because testing minimizes the defined risks, it is important that the risks are mapped out and weighed. As soon as it is known which risks jeopardize the anticipated goal the most, the test efforts can be aimed at minimizing them. This also prevents spending time and money on things that do not contribute to the goal.

- **The Generic Test Strategy: Generic Testing Guidelines**
  The generic test strategy contains the standard guidelines for all of the test projects the organization carries out. Think of the test methodology to be used, the definition of the test levels, the organization, the stakeholders and, for example, the error resolution process. Using these guidelines as the starting point means that the test plan can be less elaborate.

- **The Test Budget and Planning**
  The test budget and planning indicate how much effort the prescribed test strategy will take and when the effort will have to be made.

There is never enough time to test everything, which is why the anticipated goal, the established risks and the test depth deserve a close look when establishing the budget and the planning. Choosing a different test approach may save time, but it has to be justified by the test risk analysis: which risks are not or are only partly covered by the tests? The balance between the three components determines the test project specific test strategy. The optimal balance is determined by the result

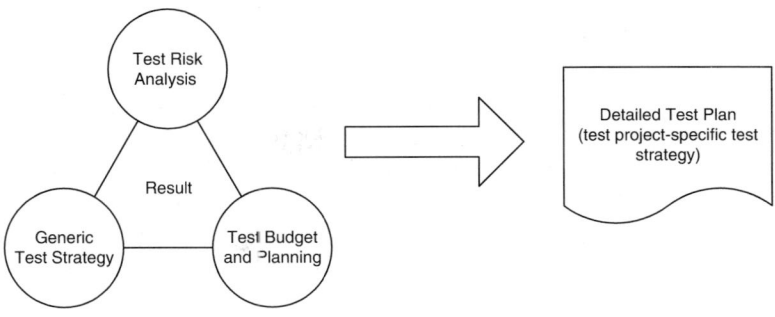

The test strategy for the test project that is described in the test plan is not autonomous, but has a relationship with the generic test strategy, the test risk analysis, and the test budget and planning. These three components are related to the anticipated goal that is described in the Goal description. The balance between these four components determines the test approach that will be used for the test project.

The Goal description was discussed in Chap. 6. The next chapters discuss the test risk analysis, the generic test strategy, the test budget and planning, and the detailed test plan, which contains the project-specific test strategy.

7 Test Risk Analysis

7.1 Introduction

Most of today's test projects are based on risks. In risk-based testing (RBT), the test approach is determined by the risk the organization runs when the system goes live. The scope and depth of the tests vary according to the risk.

The advantage of risk-based testing is that time is spent on and attention given to things that have added value for the anticipated goal [Pinkster et al, 2004]. A test risk analysis is carried out to determine the risks that jeopardize the anticipated goal. The insight provides a guideline for decisions that have to be made at various points during the test project.

The TRA is used to determine the test strategy. There is rarely enough time in a test project to test everything. The TRA can be used to set priorities, i. e. indicate what the test activities should focus on. For example, the test coordinator can use the TRA to determine which things will be tested less thoroughly or not at all. The TRA supports the choice for the test design techniques that are applied.

During reviews and smoke tests, the TRA can be used to ensure that important components will be paid more attention than less important ones. During testing, the TRA is used to establish the sequence in which the tests are run. It is common to start with the most important tests because it increases the chance of finding the important errors quickly. It also has the advantage that the most important tests will have been run should the test be stopped prematurely. The risks are referred to in the test report. The benefit of this is that the test report contains information

D.-J. de Grood, *TestGoal*,
DOI: 10.1007/978-3-540-78828-7_7, © Collis B.V., Leiden, The Netherlands, 2008

about things that appeal to the stakeholders, namely the risks for the anticipated goal.

There are two types of TRA that can be used alongside each other.

1D TRA

The one-dimensional TRA determines the relative importance of the test object's various components. A functional decomposition is used to create an overview of the functions the system should be able to support. The TRA defines which functions are important and which are less important.

The system specifications or the requirements are used as input for the 1D TRA because they describe the functions the system should support. The product of the 1D TRA is a list of these functions ranked by their relative importance. The 1D TRA can be used to determine the test depth for each function as well as the sequence of the tests.

2D TRA

The two-dimensional TRA maps out the threats to the anticipated goal. For each threat, the chance that it really will occur and what its impact would be are estimated.

The threats that were indicated by the stakeholders are used as input for the 2D TRA. The product of the 2D TRA is an "impact x probability" matrix in which each risk is positioned. The 2D TRA can be used to determine which tests have to be run. This will often result in the testing of non-functional quality attributes.

> The advantage of the 2D TRA is that it is the more accurate of the two TRAs and ties in best with the anticipated goal. The 2D TRA maps out the risks for the business and stimulates testers to look beyond the system specifications or the requirements. The test risk analysis can reveal risks that were not addressed in the system design. If the risks that were overlooked are discovered on time, they can be processed in the design before coding starts and thus prevent errors occurring early in the product life cycle.

In the testing world, risk-based testing is a household word. But there are still a number of organizations that do not do it. Experience shows that organizations have difficulty allocating the time and applying the discipline required to carry out a 2D TRA. Not only does a 2D TRA take more time than a 1D TRA, the customer might not appreciate a process that questions the system design. Whether wise or not, there are projects in which the assignment forces the system to be built according to the design. The customer uses testing to demonstrate that the system works according to the specifications and has no interest in demonstrating that the design can be improved.

In these situations the 1D TRA is often sufficient. It is relatively easy to do, is easy to understand and still provides clear starting points to differentiate test depth. The 1D TRA helps determine the sequence of the test and can be used to introduce risk-related thinking. When the organization realizes what the advantages of risk-based testing are, it can always move on to 2D TRA. In short: a 1D TRA should always be done, and if there is room and time for a 2D TRA, consider doing it as well.

Testing is the reduction or the removal of risks. The art consists of choosing a strategy that provides insight into the real risks as efficiently as possible. It is also important to be able to relate the test results to the identified risks. The tester reports the test results during and after testing. By specifying in the test report how testing has removed certain threats, he demonstrates the added value of testing and indicates the issues business management no longer has to worry about. Linking the anticipated goal to the TRA in the test report gradually increases confidence in the quality of the system. This is displayed in the below figure.

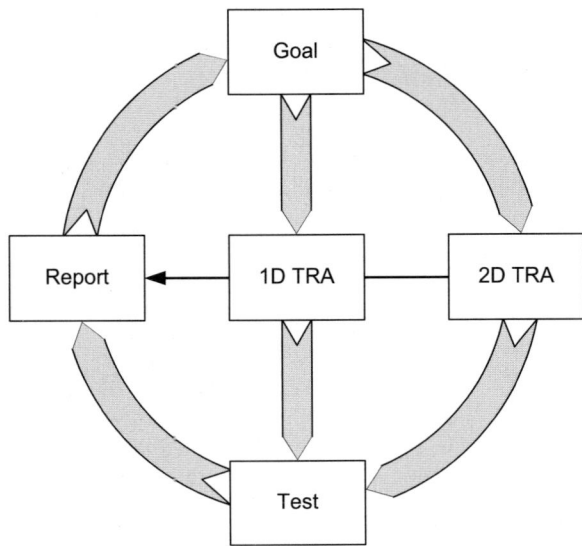

Fig. 7.1 The circle from goal to goal: The figure shows that the anticipated goal is the start and end point of the TRA, testing and the test reports

The figure shows that the anticipated goal is the starting point for all test activities. The TRA maps out which risks jeopardize the goal. The TRA is used to derive the points that need the most attention during testing. Testing itself is an activity that is carried out to examine the extent to which the risks are real. For example, if the tester is afraid that a calculation error will result in the creation of wrong invoices, he can test the calculation. He eliminates the risk by demonstrating that the calculation does not contain any errors (see the discussion on the test report). The tester relates the test results to the previously established risks, and can thus state the extent to which he thinks the anticipated goal will really be achieved.

In principle, it is not desirable to do a separate test risk analysis for each test project because it's not efficient. And because all of the activities related to the development project should be aiming for the same goal, the same test risk analysis should apply to all of the test projects. It is therefore best to do an umbrella TRA that applies to every test project. But sometimes there simply is no generic test risk analysis, and it's not possible to do an umbrella one either. This being the case, the test coordinator needs to do his own TRA.

The next section describes how a TRA is done.

7.2   The 1D Test Risk Analysis

7.2.1   Introduction

To do a one-dimensional test risk analysis, a test tree is created by decomposing the test object into functions and areas of attention. During the test risk analysis, the relative importance is determined for each branch of the tree. The result is a one-dimensional TRA matrix, i.e. a list of the risks ranked by their relative importance. See Table 7.1.

The test risk analysis is done during a workshop that is attended by various stakeholders. The stakeholders' field expertise enables them to identify the functions and areas of attention and assign them a priority. The TRA is organized by the moderator, who also guides the group through the analysis and processes the data at the end of the session. The test coordinator is often the moderator.

Table 7.1 A one-dimensional risk matrix for a navigation system

Risk category	Risk area	Relative importance
Critical	Route calculation – Standard calculation Navigation – Entering destination	270 150
High	Route calculation – Find alternative Accuracy Navigation – Favorites list	117 99 80
Medium	User friendliness Route calculation – Route type Extra – Traffic jam info Performance Navigation – Recent destination	65 63 45 45 20
Low	Navigation – Home Extra – Weather forecast Settings – Audio Settings – Maps Settings – Standard	15 9 8 6 6

The 1D TRA consists of the following steps:

1. Identify stakeholders and kick-off
2. Establish the functions and areas of attention
3. Determine the relative importance
4. Data processing
5. Agree on the TRA

The steps are explained in the next sections.

7.2.2 Identify Stakeholders and Kick-off

It is likely that the project stakeholders were identified during the assessment of the anticipated goal. This being the case, the moderator invites them to do the TRA. In the best of worlds, the stakeholders will represent a number of different disciplines. Different stakeholders probably have different visions on the anticipated goal and the risks [Thompson, 2004]. Involving various disciplines helps create a balanced and well-considered risk assessment. The group of stakeholders could be made up of, for example:

- **The Customers**
  End-users, operators, business managers or system administrators

- **The Builders**
  Analysts, system designers or programmers

Step 2 – Approach

- **The Project Owners**
  Project managers or customer

After the participants of the TRA have been selected and invited, it's time to kick off the project. The moderator explains why the TRA is done, how it works and what is expected from the stakeholders.

It is important that each party that is involved in the TRA is deemed competent by the group he represents. This ensures that the participant can make well-founded statements about the priority and that his assessment will be accepted by his group. This may seem to be a trivial issue, but it isn't. Consider the following case:

> *Example 7.1: A Web-based application*
>
> The organization is building a Web-based application that enables customers to place orders online. During the TRA, the project leader indicates that the project has a rather short runtime. As a result, more concessions will have to be made to the delivered functionality. The participants agree that the part the customers will be seeing cannot be compromised on. The Web interface must be developed and tested with diligence. But the application has another interface, namely the screens the employees will be using to process the orders. During the TRA, it is agreed that the screens have to work but that the user friendliness does not have to be tested.
>
> The user representative knows that his group will not be happy with this decision. The fact that this interface will not have been thoroughly tested may impact their work. Nonetheless, he agrees with the decision because it seems to be the wisest and he knows the users will respect his decision.

7.2.3   Determine the Functions and Areas of Attention

Before the moderator can assign priorities to the risk areas, he has to map them out. This he does by creating a test tree. A test tree is a kind of mind map in which each function or area of attention is a branch in the tree. If necessary, functions and areas of attention can also have branches, meaning that one branch can be split up into multiple branches.

The below figure shows the test tree for a simple navigation system. A number of main functions can be recognized: Navigation, Route calculation, Extra and Settings. Each of the main functions, also called function groups, is a collection of functions. The main function, Navigation, contains all of the ways in which a trip's final destination can be entered. As can be seen in the test tree, this can be done in several ways: by entering a new destination or by selecting a destination from the list of recent or favorite destinations. It is also possible to be guided home. The left side of the test tree displays some non-functional areas of attention that can also be branched out. Accuracy, for example, can be branched out into the calculation of the trip time, the timely indication of an exit or, not unimportant, the correct indication of the fastest route.

Using a test tree as the basis for the test risk analysis has a number of advantages. During the TRA, the test tree can be used as a checklist to ensure that no functions are forgotten and later as a framework for the physical test design. This also makes it easy to relate the test results to the test risk analysis and report them back to the stakeholders.

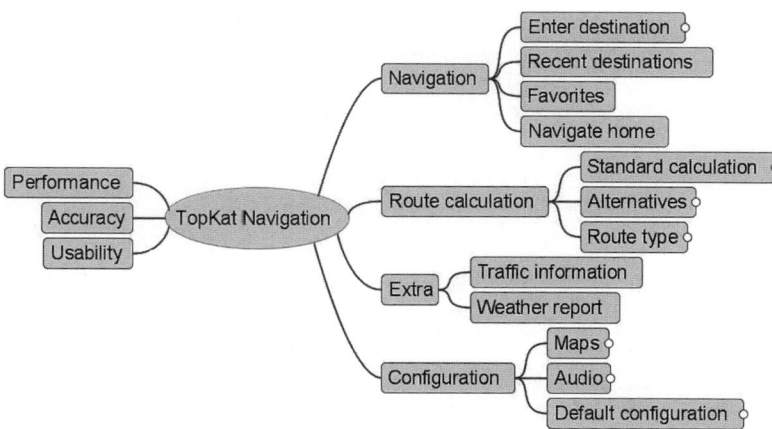

Fig. 7.2 Test tree for a simple navigation system

The functions in the test tree are derived from the test base (for example, functions described in the requirements, the functional design and use cases). The areas of attention are mentioned by the stakeholders and originate in the

- business
- processes
- functionality and technique
- quality attributes

The areas of attention are mapped out during interviews with the stakeholders, which can take place prior to or during the TRA session. Good preparation ensures that the TRA session doesn't take too long. During the TRA session, the participants check the completeness of the test tree and add new insights if necessary. Once the functions and areas of attention have been mapped out, it's time to determine the relative importance of the risk areas.

> As preparation for the interviews with the stakeholders, it can be useful to create a context diagram for the test object. The analyst may even have already created one. A context diagram is usually a kind of architectural diagram and defines the relationships between various functional components and the system interfaces. The diagram is reused in the test plan (see Chap. 10).

This diagram can be used to ask questions about the processes, information streams, functionality, security aspects, etc. Such a diagram often helps clarify things because it quickly provides insight into the functions and areas of attention. Moreover, running through the quality attributes in the Quint model ensures that the non-functional aspects are not forgotten either.

7.2.4   *Determine the Relative Importance*

An efficient approach is to ask all of the stakeholders to assign priorities to all of the branches in the test tree. Each of the stakeholders divides their points over the branches. This can be done either in individual interviews, or if the group is not too big, together. The ratings are compiled after everyone has provided their input.

The following ratings are used:

9 points
The function or the area of attention is key to the working of the system and to achieving the anticipated goal. Errors in this function or a poor implementation will directly impact the system's usability. The system can only be released after this function or area of attention has been thoroughly tested.

5 points

The function is important. Errors in these functions, or a poor implementation, are allowed if a workaround is available. Functions and areas of attention should be tested well before they can be released.

3 points

Non-crucial function or area of attention. Errors in the function can be a hindrance, but are not expected to jeopardize the anticipated goal. Functions and areas of attention must be tested before they can be released.

1 point

Function or area of attention that is not necessary for the working of the system. It is preferable to test them, but they can be tested less thoroughly.

When awarding points, it is important to aim for a good division over the risk areas. Participants who are not very experienced in estimating risks will be inclined to find everything important. In this case, stress that it is important to make distinctions. Explain the purpose of the test risk analysis again and stress that it is about relative importance. Everything may indeed be important, but even then some things are more important than others. Table 7.2 displays the data of a TRA session with four participants.

Table 7.2 Data from a TRA session

Risk area	Spread 1	Spread 2	Spread 3	Spread 4	Total
Navigation-Entering destination	9	9	3	9	30
Navigation-Recent destination			3	1	4
Navigation-Favorites list	9	1	3	3	16
Navigation-Home			3		3
Route calculation-Standard calculation	9	9	3	9	30
Route calculation-Find alternative		5	3	5	13
Route calculation-Route type		1	3	3	7
Extra-Traffic jam info		3	3	9	15
Extra-Weather forecast			3		3
Settings-Maps			3	3	6
Settings-Audio		5	3		8
Settings-Standard		3	3		6
Performance	9	3	3		15
Accuracy		5	3	3	11
User friendliness	9	1	3		13
Total	45	45	45	45	

Step 2 – Approach

Should an attempt to distinguish between the functions and areas of attention fail, give the stakeholders a limited number of points. This was done in the above example. The participants had divided a maximum of 45 points over 12 risk areas. Only 5 x 9 points can be awarded (Division 1). If the participant wants to give points to more than five risk areas, he is forced to give fewer points (Division 2). If a participant wants to divide his points evenly, that's fine too (Division 3). However, because there is a maximum to the number of points that can be given, he minimizes his overall influence. If this is pointed out to the participant, he will probably decide to divide his points differently.

It is very well possible that a function that is not very important on its own contains a number of paths that are important. This becomes clear if not only the lowest level of the test tree is assigned a relative importance, but the level above it as well. In the below example, a priority has been assigned to each function and subfunction. The functions and subfunctions have been assigned their priorities independently of each other. When they are multiplied by each other, it is striking that the relative importance of the subfunctions get mixed up. In the below table, the relative importance of both functions has been taken into account in the total calculation.

What is striking is that the function Extra is found to be much less important than the function Navigation, which get 3 and 5 points respectively. Yet, the relative importance of *Extra-Traffic jam info* is higher than that of *Navigation-Recent destination*. When doing the TRA, take into account that a less important function can contain important aspects.

Table 7.3 The relative importance of functions

Risk area	Relative importance of function	Relative importance of subfunction	Total
Route calculation-Standard calculation	9	30	270
Navigation-Enter destination	5	30	150
Route calculation-Find alternative	9	13	117
Accuracy	9	11	99
Navigation-Favorites list	5	16	80
User friendliness	5	13	65
Route calculation-Route type	9	7	63
Extra-Traffic jam info	3	15	45
Performance	3	15	45
Navigation-Recent destination	5	4	20
Navigation-Home	5	3	15
Extra-Weather forecast	3	3	9
Settings-Audio	1	8	8
Settings-Maps	1	6	6
Settings-Standard	1	6	6

7.2.5 Process the Data

Each of the participant's data is gathered and sorted by relative importance. This can be done in Excel.

The risk areas are then grouped into risk categories, which are the product of the test risk analysis. Table 7.4 shows how the risk areas can be divided over the risk categories.

Table 7.4 Risk categories

Risk category	Content
Critical	Most important 10% of the risk areas
High	Next 20% of the risk areas
Medium	Next 30% of the risk areas
Low	Least important 40% of the risk areas

Step 2 – Approach

Using the example in the previous section, this produces the following TRA:

On <date> a risk analysis was done by:
Name, job title
Name, job title

The following risks and relative importances were identified:

Risk category	Risk area	Relative importance
Critical	Route calculation – Standard calculation	270
	Navigation – Entering destination	150
High	Route calculation – Find alternative	117
	Accuracy	99
	Navigation – Favorites list	80
Medium	User friendliness	65
	Route calculation – Route type	63
	Extra – Traffic jam info	45
	Performance	45
	Navigation – Recent destination	20
Low	Navigation – Home	15
	Extra – Weather forecast	9
	Settings – Audio	8
	Settings – Maps	6
	Settings – Standard	6

Fig. 7.3 An example of a TRA

7.2.6   Agree on the TRA

The last step consists of agreeing on the TRA. Once the estimates have been processed and the risk categories indicated, it is wise to ask the participants and stakeholders for their approval. Ask them to make sure that no areas of attention have an illogically high or low position in the TRA. At this stage, remarks can be processed without any problems. It is better to process new insights sooner rather than later.

> Also ask the stakeholders to approve the test tree. Areas of attention may have been added or removed during the TRA. The approved test tree is used for test clustering during the test design phase and is the basis for the test reports.

Once the TRA has been approved, the test strategy can be selected. The test strategy is described in the next chapters.

7.3   The 2D Test Risk Analysis

7.3.1   Introduction

Risks are identified differently in the 2D TRA than in the 1D TRA. The main difference between the two is that the 1D TRA is based on the system specifications or requirements and the 2D TRA is not: the 2D TRA is based on the risks that jeopardize the anticipated goal. The 2D TRA thus enables risks to be detected that are not addressed in the test base.

Once the anticipated goal has been defined in the Goal description, the threats to the anticipated goal are examined. A threat is any event that can prevent the goal being achieved. Because not all threats need to be taken equally seriously, the risk of each event is estimated. The following definition of risk is used:

> The risk of an event is the product of the chance that the event will occur and the impact it has on the goal. Or: Risk = Chance x Impact.

The goal is included in this definition. In result-driven testing, we refer the identified risks back to their effect on the anticipated goal. This is explained using the Connecta project (see Example 1.3).

Example 7.2: Risk and goal at Connecta

The anticipated goal is clearly formulated in the Goal description for Connecta's system test. Business is counting on the system using fewer resources. Integrating a number of existing systems not only accelerates the processes, it also reduces the number of errors. If we take this anticipated goal as a starting point, we immediately identify a risk. What happens if the new system doesn't reduce the number of resources? This could be the case if the new integrated system does not:

- process data faster, but just as fast or slower
- produce less errors

The tests will have to make a statement about the degree to which the project's goals can be achieved. This is why the test coordinator will at least have to ensure that the tests make a statement about the speed at which the system works and the errors that occur.

Each of these risks can be analyzed more closely. The aim is to find out which events trigger the stated risk. An Ishikawa diagram, also known as fishbone diagram, is often used for this (the diagram is referred to as fishbone diagram because of its shape) [Bilt]. Figure 7.4 displays a fishbone diagram.

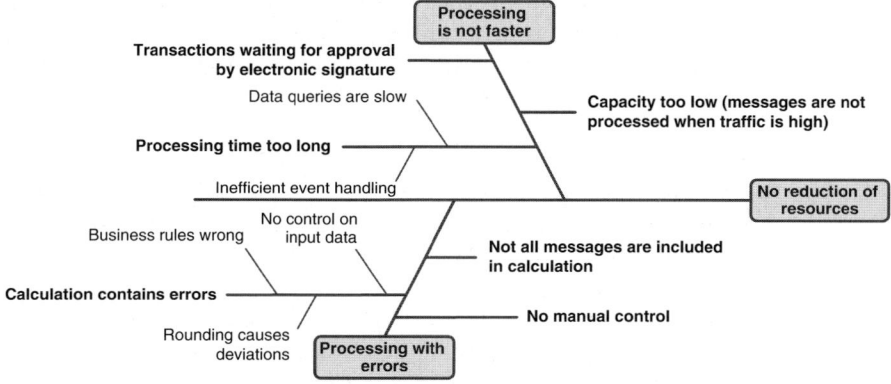

Fig. 7.4 A fishbone diagram

In a fishbone diagram, causes and effects are linked to each other. In this case, the diagram shows events that present a risk. If, for example, not all messages are processed, the output of the process will probably contain errors. If the process contains errors, the organization will have to make manual corrections. This is why saving resources was estimated as a risk.

After the risks have been mapped out, measures can be taken to deal with them. In practice, this means that one of the following options will be chosen:

- Not tackle the risk
- Study the extent to which the risk is real
- Remove the causes that lead to the risk
- Reduce the effects of the risk

Testing is mainly aimed at the two points in the middle. Testing enables experience with the working of the system to be gained as well as insight into the actual chance of the risk occurring and its potential impact. Finding and solving errors removes the cause of the risk.

Example 7.3: Covering the risks in the Connecta project

Among others, the following two risks were identified in the Connecta project:

Risk 1:   An error in the calculation causes a lot of manual follow-up.
Risk 2:   The capacity of the system is insufficient and cannot process all of the requests on time.

The test project aims at covering these risks as follows:

Risk 1:
The calculation is tested with realistic scenarios. The outcome of each step is compared to the expected outcome. In addition, illogical combinations of data are entered to check if error handling is working properly. During the test, a number of errors were found that caused the calculation to produce the wrong outcome. By solving these errors, the chance that wrong calculations are made in the live system is considerably reduced. The test result is discussed in the test report. The stakeholders believe that risk 1 has been reduced to an acceptable level.

Risk 2:
The organization knows that a lot of requests are received every day. Using this load, a performance test is carried out to determine the peak load the system can bear. At a high peak load, messages are sometimes not processed. Research, however, reveals that it is not expected that this peak will occur in the coming years. Subsequently, an endurance test is run to demonstrate that if the load is 150% of the current number of requests, the system will run without problems for one week. The test shows that risk 2 is still present but does not currently require attention.

The test report provides information about the identified risks and their status. By referring the test results to the previously defined risks,

Step 2 – Approach

statements can be made about the extent to which the risks still jeopardize the goal.

The 2D TRA consists of the following steps:

7.3.2   *Identify Stakeholders and Kick-off*

The TRA participants are selected, invited and informed in the same way as for the 1D TRA (see Sect. 7.2.2).

7.3.3   *Establish the Risks*

The TRA participants are asked to

- point out risks
- examine the causes a risk can have
- mention factors that can influence the chance of failure

For the 2D TRA, the chance and the impact are estimated separately for each of the established threats. This is done by assigning points to each factor for both the chance of failure and the impact. Estimating the impact enables the possible damage the risk will incur to be determined if it occurs. The possible damage relates to the anticipated goal and can be subdivided into, for example:

- Finances
- Image
- Internal business processes
- External business processes
- Business processes of customers, direct or other

When estimating the chance of failure, it is also estimated how high the chance is that the risk will actually occur. The chance depends on the following factors:

- The complexity of the test object's component
- The size of the test object's component
- The extent to which new unproven technologies are used
- The relationship with other components
- The quality of the build team
- The frequency with which the component is used

7.3.4 Data Processing

After all of the risk workshop participants have handed in their input, the impact and chance of failure are calculated for each risk. The results are commonly presented in a chance-impact graphic. Categories are assigned to the risks in the same way as for the 1D TRA.

Risks with a high chance and impact are critical. The risks with a low impact and chance belong to the low risk category. The below graphic displays the risk categories.

The TRA matrix is often created with four quadrants (Fig. B). The "critical" risk quadrant contains all of the risks for which the chance of failure and the impact are high. However, this representation is not recommended. The division of the quadrants does not correspond to the distribution of the risk in the graphic and can cause confusion. Graphic C shows the product of chance and impact in each cell. The graphic shows that the upper right corner of quadrant 3 represents more risk than the lower left corner of quadrant 1. The cells have a value of 27 and 25 respectively. Graphic A displays a more realistic image of the risk categories. In graphic A, the points with the same risks have been connected with each other. The risk categories correspond to the distribution indicated in Graphic C [Baars et al, 2006], [Gardiner, 2006], [Pinkster et al, 2004].

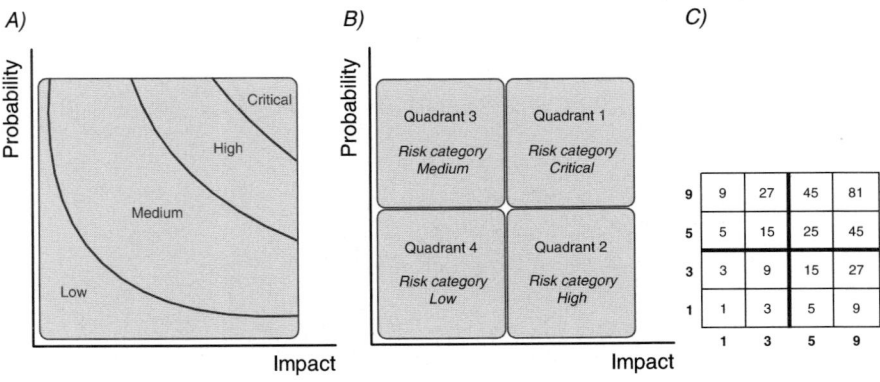

Fig. 7.5 The TRA matrix. The risk of an area of attention, expressed in estimated failure chance and impact, can be represented by a point in the chance-impact graph. A) Realistic distribution of risk categories, B) Distribution in quadrants gives a distorted picture of the risk categories, C) Distribution of risk in the TRA matrix.

A list such as the one we saw in Sect. 7.2.1 can be extracted from the graphic. This is in fact a list of the risk areas sorted by relative importance.

7.3.5   *Agree on the TRA*

As for the 1D TRA, it is wise to ask the stakeholders for approval after all of the data has been processed. See also the description in Sect. 7.2.6. Once the TRA has been approved, the test strategy can be chosen. This is described in the next chapters.

8   Generic Test Strategy

Step 2 – Approach

8.1   Introduction

Why is an organization prepared to spend time and money on test activities? Testing makes a statement about how realistic it is that the anticipated goal will be achieved. It points out the risks that jeopardize the
goal and studies the impact of changes that are implemented. The test
process reduces the uncertainties and contributes to the confidence in
the built system. Moreover, the errors found during testing can help improve the quality of the system.

Now that the goal of testing is clear, the following challenge is waiting:
how is this goal going to be achieved? This is where the test strategy
comes in: it describes the plans to achieve a goal [Kramers, 1987].

The test strategy can be used to communicate with others about the set
up of the tests and the strategic choices that have to be made. The test
strategy explains how the anticipated goal is translated into the manner
of testing. To arrive at the test strategy, the organization's existing
guidelines, the test risk analysis, and the planning are combined into
one consistent whole. The test strategy can be added to a number of test
products, such as:

- The generic test strategy
- The master test plan
- The detailed test plan

The contents of the test strategy are more or less the same, wherever it's
mentioned. In a generic test strategy, the strategy will be less detailed
than in a detailed test plan. But the test strategy has the same goal in each
of the cases: to determine how the anticipated goal will be achieved.

D.-J. de Grood, *TestGoal*,
DOI: 10.1007/978-3-540-78828-7_8, © Collis B.V., Leiden, The Netherlands, 2008

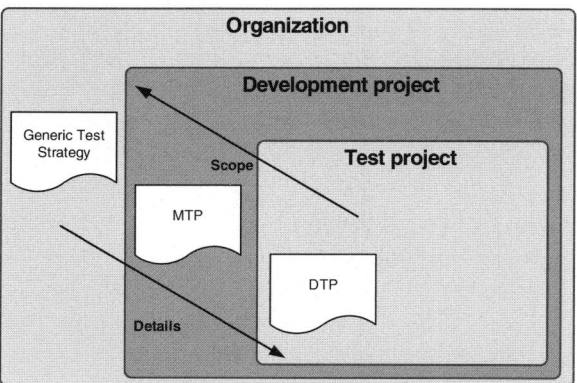

The test strategy can be described in a number of different documents. The generic test strategy describes the things that apply to the whole organization. The MTP describes in more detail how the generic document is applied to a specific software development project. The DTP has a more limited scope: it specifies what the approach is for a specific test project and it specifies in detail where the strategy differs from the strategy described in the MTP.

8.2   The Generic Test Strategy

In an organization where a lot of software is developed, a generic test strategy can considerably increase efficiency because it defines a generic approach that applies to every test project. A generic approach makes test projects

- More efficient; the wheel does not have to be constantly reinvented
- Easier to compare; each project uses the same starting point and the same definitions
- Easier to control; a generic test strategy makes it possible to define fixed control points and use them to monitor the project

Among other things, the generic test strategy describes the following items:

The Importance the Organization Attaches to Testing
The generic test strategy finds its right to exist in the realization that testing is important for the organization. The generic test strategy explains why the organization thinks testing is important.

Example 8.1: The importance of testing

"In our marketing statements, our company emphasizes that we supply reliable and high-quality products. To be able to maintain this product positioning, we have to ensure that our products are indeed reliable and of high quality. This is why controls are carried out during product development, maintenance and delivery. Testing is one of these controls and focuses on the development and maintenance of the software in our products."

The Generic Areas of Attention and Risks

For a product line or range of services, generic areas of attention and risks can often be defined. Although there can be additional project-specific areas of attention, the generic areas of attention and risks always apply. The above example mentions a number of generic areas of attention. Reliability is an important property for all of the products the company supplies. Other examples are security for banking systems: who wants to leave their money in a bank that is said to be unsafe? Or continuity for an airline's flight booking and reservation systems: if the site is not accessible, the traveler will book his flight elsewhere.

These areas of attention mention things that are important if we want to achieve the business goal. The generic test strategy describes the areas of attention and what should be done with them. The strategy shows how it is guaranteed that these areas of attention will actually be tested and that the focus remains on the anticipated goal. This can be done, for example, by grouping them in a specific test level.

A Definition of Test Levels and Notions

A definition of the test levels that are used in the organization ensures that all of the parties have the same notion of their contents. It helps the stakeholders aim for the same goal and prevents wrong expectations. Section 4.2 describes a number of frequently occurring test levels such as the user acceptance test (UAT).

User Acceptance Test

The user acceptance test is mainly a validation test (is the system "fit for purpose"). The tests are based on representative scenarios from the users' daily jobs. The test is run to determine whether the users can work with the system, how usable the system is and how the system integrates with the working method and processes.

Step 2 – Approach

A generic glossary can prevent discussions and confusion arising about the used terminology, and ensures that we can focus on the things that deserve our attention. Appendix E contains a glossary with the terms that are used in TestGoal.

The Existing Test Environments
The generic strategy describes the exiting test environments and their characteristics, and indicates which test levels are run in which environment. The description in the generic test strategy is usually generic to prevent having to adapt the strategy. The environment is described in more detail in the test plan. See Chap. 10 Test Plan.

The Relationship with the Organization's Processes
General processes, such as error management, change management and release management, and the tools used can be described in the generic test strategy.

Test Risk Categories and Test Design Techniques to be Used
The classification of the risk categories and a description of the test design techniques that can be used ensure that the various test projects are set up generically.

Organization
The generic test strategy defines the parties that are involved in the test project. Because it is generic, the generic strategy does not mention the names of the people involved, but their job titles and the associated responsibilities. If, for example, an MTP says that David Bloom is the test coordinator for the system tests, everyone knows what David is responsible for. A description of test functions can be found in Chap. 3. Moreover, the different departments can specify their involvement in the test process and their responsibilities. Everyone knows that the development department plays a role, but does this also apply to the marketing and QA departments and the helpdesk?

Control
The generic test strategy also describes how the testware should be controlled. This component ensures that clear agreements are made about the maintenance of the test set. This is important for regression tests.

A generic test strategy works well in organizations that have a dedicated testing department. A generic test strategy also has added value if a lot of software development projects are carried out and the organization believes that consistent quality is important. The added value is also high for tests that consist of standard test activities, such as conformance tests.

8.3 Test Strategy in the DTP and MTP

If available, the generic test strategy should be the starting point when creating the MTP or DTP. Generic agreements that apply to a project are also easy to reuse. There may be cases in which more detail is needed or a generic scenario does not work for specific points. The test plan specifies which points diverge from the standard by

- Referring to the MTP and the generic test strategy in the DTP, and
- Referring to the generic test strategy in the MTP

This is done for a number of reasons:

- It keeps the test plans short because they only contain the non-standard aspects of the test project, which are also the most interesting.

- Test projects are not autonomous; after all, the anticipated goal and the associated risks apply to the entire system. The test manager ensures that the test projects tie in with each other and cover all of the relevant risks and areas of attention. This means that the test strategy should be determined at a higher level. An individual test project targets a subset of these aspects.

- Reinventing the wheel for each test project is inefficient. Think of the creation of processes such as release management, error management and the maintenance of the test environment. This is why such things are done on a project or organizational level.

To ensure that all of the test levels are efficiently and consistently set up, the generic parts of the test strategy are included in the MTP or the generic test strategy. The lower-level detailed test plans can refer to them. The detailed test plan indicates how direction can be given to the path specified in the MTP. The plan also shows where the test project diverges from the prescribed approach. Chapter 10 Test Plan discusses the test strategy in detail and mentions the things that can be included in a strategy.

> **Cutting and Pasting**
>
> Rule of thumb: When a piece of text about the test strategy is copied for the second time from a detailed test plan for reuse in a different detailed test plan, it is worth considering whether this text should be included in the MTP or the generic test strategy.

9 Test Budget and Planning

9.1 Introduction

This chapter helps you create a test budget and planning according to the TestGoal step plan. The advantage of doing this is that the activities and products of the test project have already been defined in the step plan. This ready-made list of activities and products accelerates the creation of the budget. Hours that are often forgotten are embedded in the step plan and automatically find their way into the budget and the planning. This chapter contains guidelines for checking the budget and making the planning in order to produce a realistic and well-founded planning.

The anticipated goal is described in the Goal description. The budget is an estimate of the costs that have to be made in order to achieve the anticipated goal. The test budget consists of an estimate of fixed and variable costs and is in principle time independent. The budget identifies the necessary activities and requisites, but does not indicate when they have to be completed. The planning takes the time component into account. This produces a phasing that indicates the sequence in which the test activities should be carried out, and defines when the milestones should be reached.

In general, experience data is the ideal basis for creating a budget. If this data is not available, the budget has to be created in a different way. There are various techniques to estimate the amount of work. In this chapter, the work breakdown structure (WBS) is used; this estimation method can be used with and without experience data and has the advantage that it demands a structured approach. The WBS is a good starting point for a detailed planning and a structured test report.

D.-J. de Grood, *TestGoal*,
DOI: 10.1007/978-3-540-78828-7_9, © Collis B.V., Leiden, The Netherlands, 2008

The following steps explain how the test budget and planning are created.

- Test budget
  - Determine the anticipated goal
  - Work breakdown structure (WBS)
  - Assess the requisites
  - Determine the budget
- Test planning
  - Generic planning
  - Detailed planning

The checklist helps create a test budget and planning that are as complete as possible.

9.2   Create the Test Budget

9.2.1   *General*

A budget consists of the following components:

Variable Costs
Variable costs directly depend on the amount of work or the number of employees. The largest variable cost items usually consist of the hours of internal and external employees, but the rent for the workplaces and the technical support costs can also fall under variable costs. Software licenses that are issued by user or period are also considered as variable costs.

Fixed Costs
Fixed costs do not depend on the amount of work or the runtime of the test project. This category contains, for example, the one-time costs that are made for hardware and the software licenses that are part of the test environment or the required infrastructure.

Many projects only take the variable costs into account, for example, when the test environment and licenses are already present. Before creating the budget, find out which costs it should include. If the customer is only interested in the variable costs, then use them as the starting point. The majority of the variable costs consist of the employees' activities. This is why the activities that need to be carried out are first defined using a WBS.

9.2.2 Work Breakdown Structure

A WBS is created by breaking down the test project into small chunks. Each piece consists of the work that has to be done to complete an activity or produce a product. The main activities of the WBS are mapped out first. Then, for each main activity it is determined what is needed to achieve it. This produces a list of subactivities. Figure 9.1 displays a WBS for drinking of a cup of tea.

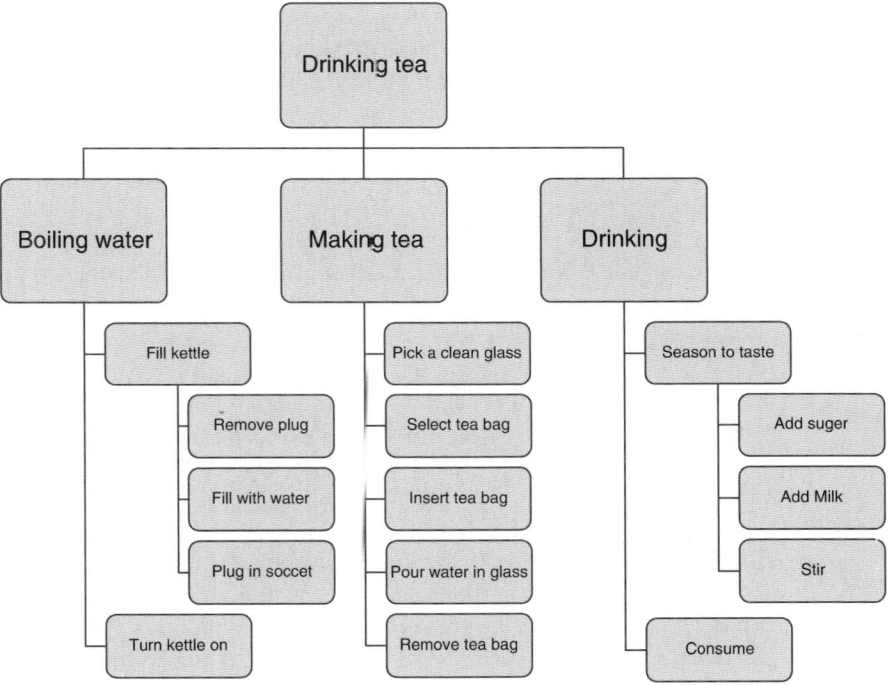

Fig. 9.1 A WBS for drinking a cup of tea

The WBS provides insight into the main activities. In this example, they are Boiling water, Making tea and, ultimately, Drinking tea [Business-Future]. By defining the main activities, a number of milestones have been determined as well. When the boiling water phase is finished, boiling water is available. When the making tea phase is finished, there is a glass of tea, which is ready for consumption, but may have to be seasoned to taste. When the drinking phase is finished, all that remains is the empty glass.

The WBS also indicates how a main activity branches out into subactivities. In the example, the kettle's plug is taken out of the socket for safety reasons before the kettle is held under the faucet. If sugar is added, stirring is necessary.

The elaboration of a WBS often raises many questions, and their answers often provide a lot of insight into the set up of the test project. Discuss the possible details with others involved in the test project. This reduces the number of surprises that arise later in the project, leads to a balanced phasing and ensures the approach is supported.

A closer look at the above example reveals that the WBS is not yet complete. To stir, you need a spoon. Getting the spoon can be included in the WBS as a subactivity. The first instinct may be to put this subactivity under making tea. By doing this, the milestone of this activity remains standing: the tea is ready to be consumed. On the other hand, if the user does not want sugar, he does not need a spoon, so maybe this subactivity should be put under drinking tea. Both choices are right, but have different consequences.

Something else is missing: If a clean glass is taken from the cupboard, it may have to be washed after drinking. This activity also takes time and costs money and should hence also be included in the budget.

Making a WBS does not only help create a good budget, is also helps define and limit the activities, which is why the WBS is a good basis for the test report. The extent to which the goal has been achieved can be derived from an activity's progress.

The level of detail in the WBS varies depending on the size of the project. After all, a detailed list of activities is easy to create for a small project. For large projects, however, activities are combined to maintain the overview. How detailed should the WBS get? A rule of thumb is not to include any activities that take less than a day to complete, with the exception of critical activities that drive the test project, such as milestones; an example of this is "Acceptance (test plan ready)." In the planning, these critical activities are called milestones, and to prevent them being forgotten in the planning, they are included in the WBS.

The following steps apply to creating a budget that is based on the WBS:

- Establish the anticipated goal
- Assess activities and products
- Estimate the number of hours required for each activity
- Estimate the costs
- Control

Determine the Anticipated Goal

The budget is an estimate of the costs that have to be incurred in order to achieve the anticipated goal. The anticipated goal is determined during the assessment of the anticipated goal and is described in the Goal description.

Assess Activities and Products

In order to create a testing budget, a WBS is created for the test project, which maps out all of the products and activities that have to be carried out. The WBS is based on the TestGoal step plan, in which all of the important activities are described. Basing the WBS on the step plan has the advantage that essential activities are not forgotten. The products and activities in the step plan are listed in the below checklist.

Table 9.1 A checklist for the products and activities in the step plan

Test activity	Subactivity
Goal	
Assess anticipated goal	• Initial meeting • Formulate the goal description • Review and adaptation • Acceptance (Goal description ready)
Approach	
Test risk analysis (TRA)	• Identify stakeholders • Prepare TRA meeting • TRA meeting • Review and adaptation • Acceptance (TRA ready)
Budget and planning test assignment	• Frame the budget • Review and adaptation • Acceptance (Budget ready) • Create generic planning • Create detailed planning • Review and adaptation • Acceptance (Planning ready)
Determine test strategy	• Frame and discuss initial strategy • Determine strategy • Review and adaptation • Acceptance (Strategy ready)
Create test plan	• Create test plan • Review and adaptation • Acceptance (Test plan ready) • Kick off approach and methodology

Step 2 – Approach

Table 9.1 Continued

Test activity	Subactivity

Design

When planning the sanity check on the test base and the test design, the test tree we created for the test risk analysis provides a lot of detailed information. The test tree indicates which tests have to be defined. When scheduling the design activities, an estimate of the required time can be specified for each branch in the test tree. The one-to-one relationship with the TRA enables the thoroughness of the tests to be taken into account. As a result, relatively more time is scheduled for the important components. The outcome of the TRA can also be taken into account for the sanity check. This enables more time to be spent on the test base components that describe an important function.

Test activity	Subactivity
Sanity check on test base	• Determine and compile test base • Plan and run sanity check • Reporting and consultation
Test design	• Draw up logical test design • Review and adaptation • Draw up physical test design • Create test scenarios • Review and adapt test design and scenarios • Define test data • Adapt test design according to changes in the test base Estimate the number of times the test design will have to be adapted as a result of changes to the test base. Estimate whether the nature of the changes entails a repetition of the sanity check.
Requirements for the test environment	• Assess requirements for the test environment • Draw up requirements for the test environment • Discuss with maintenance organization • Review and adaptation • Acceptance (requirements for the test environment ready) • Purchase requisites

Set up (Iterative process)

Estimate the number of times an environment has to be set up and how many test runs are required to reach a positive release advice.
Experience data is often available when testing in a maintenance organization. Use the data to estimate how often an environment has to be set up.

For new developments, experience data is generally not available. In this case, assume that three test runs are needed and that the environment will have to be set up at least as many times. Assuming fewer test runs is only realistic if the test object is of high quality upon initial delivery. More iterations are usually not necessary unless it can be demonstrated in advance that the quality of the test object is very low and the anticipated goal is seriously jeopardized. Also see Chap. 18 Test Execution.

Test activity	Subactivity
Set up the test environment	• Set up the infrastructure • Set up the test environment
Configure the system	• Install the test object and test data • Configure the test object/test environment
Smoke test	• Plan and run smoke test • Reporting and consultation • Rerun tests (showstoppers)

Table 9.1 Continued

Test activity	Subactivity
Execution (Iterative process)	
Run tests	• Run defined tests • Rerun tests (non-showstoppers) • Run regression tests • Record errors
Release	• Gather test results • Test against exit criteria • Draw up final release advice • Draw up final test report
Assurance	
Determine regression test	• Determine regression test
Evaluate test project	• Plan and carry out evaluation • Write lessons learned report • Review and adaptation • Acceptance (lessons learned report ready)
Archive testware	• Assess destination of software • Archive testware and errors • Clean up project directory
Transfer	• Transfer to stakeholders
Discharge test team	• Discharge test team
Miscellaneous	
Coordination and management	• Test coordination (guide testers) • Test team meetings • Test reporting
Consultation	• Project meetings • Triage meetings • Review meetings • Workshops • Brainstorming session
Training and introduction	• Testing training • Training/introduction of company and/or system • Introduce employees to the work in terms of the method/system
Control	• Set up maintenance of test environment and data • Maintenance on test environment (during test project) • Maintenance on test data (during test project)
Unforeseen	• Unforeseen (10–15% of the total estimate)

Hour Estimation

The WBS contains an estimate of the number of hours required to complete each activity. The following is needed in order to create a realistic planning:

- **Provide an Estimate of All of the Activities**
  Starting with a complete WBS prevents hours being forgotten. Experience shows that the time spent on reviews, processing comments, rerunning tests, training and meetings, is often forgotten. These points are included in the checklist.

- **Estimate Accurately**
  - Including enough detail in the WBS ensures the activities can be estimated with reasonable certainty. Errors will, of course, be made, but because the estimates are made for each activity, the deviation will be small and maybe even compensated by deviations in other estimates.
  - Estimating an activity accurately seems to a very difficult thing to do. Experience helps, but it still is more of a gut feeling than an exact science, which can make the estimate difficult to support. Metrics provide a good foundation and make it easier to adapt the budget if, for example, it turns out that one of the assumptions is incorrect or if the composition of the release changes. Metrics would help explain the amount of additional time that is required or can be saved if, for example, two use cases were removed from the release.

Examples of simple metrics are:

- **Review**
  Metric: number of pages per hour.
  The estimated time required for the review based on the number of pages in the test base.

- **Test Design**
  Metric: Number of test cases per use case, Number of test cases per hour. The estimated time for the test design is the Number of use cases x Number of test cases per use case / Number of test cases per hour.

- **Test Execution**
  Metric: Number of test cases per use case, Number of executed test cases per hour. The estimated time required to run the test is the Number of use cases x Number of test cases per use case / Number of test cases per hour.

If no experience data is available, a "best guess" is made when framing the budget. As soon as an activity starts, the number of pages actually reviewed per hour can be controlled, or the amount of time actually required to create a test case. This new knowledge can be used to adapt the metrics, and the adapted metrics can be used to guess again how much time is required to complete the activity. The remaining required time is also referred to as the estimated time to completion (ETC).

If there is enough time and the test base is available, it can be useful to review part of it with the team, or to elaborate a representative part of the system design. This baseline measurement forms a basis for other metrics.

The total number of estimated hours is obtained by adding up all of the required hours.

9.2.2.1 Cost Estimate

The budget is created by converting the hours into costs using the hourly rate in one of two ways.

Use an Average Hourly Rate
This option provides a quick result because no detailed planning is required and the people who will be working on the project do not have to be known. However, watch out for differences that can arise if the actual rates are used in the test report or to settle accounts.

Link the Activities to Employees and Calculate the Costs According to their Hourly Rate
This option is more accurate, but also more complicated. For internal employees, the hourly rate is the internal labor rate. For external employees, it's the contracting rate.

In order to do this, the availability and the rates of the employees that are used in the calculation must be taken into account. If the scheduled employee is not available and the activity is carried out by a different employee, the activity may become more or less expensive. To avoid surprises, a detailed planning is necessary to establish which employee is available when. Creating such a detailed planning takes a lot of time and may be a bit premature if the budget has not been formally approved. This is done at a later stage, namely after the planning has been approved and if the customer declares that the first option is not accurate enough.

9.2.2.2 Verification

A number of controls can be carried out to check whether the estimate is correct. They can be used individually or together.

Peer Review

Let a colleague who has experience with budgeting and testing check the estimate. Focus the review on the completeness of the activity list and on the correctness of the estimate.

Independent Estimate

Let a colleague make his own estimate. Discuss the differences and the total amount of the estimate. Adapt the estimate in such a way that there is consensus.

Experience Data

Evaluate the estimate against earlier estimates. Compare the size and complexity of the reference project to the current project, and convert the hour planning to the current estimate. Bear in mind that the reference project may not have been set up according to the planning. It may be preferable to use the actual number of hours spent as the starting point rather than the planned hours.

The size and complexity of the projects can be compared in a number of ways, for example, by doing a function or test point analysis.

Statistics

A variety of statistic calculations for the relationship between development costs and testing costs, or between the phases of the test project are known from literature.

Jones [Jones, 2000] indicates for each test type what the effort of the tests is in relation to the total project effort. Because the hours that are spent on the module tests are often included in the development costs, they are not included in the total test effort. According to the below overview, the relative test effort corresponds to the frequently rule of thumb norm that testing constitutes 40 percent of the project budget.

Table 9.2 The test effort in relation to the total project effort. Source: [Jones, 2000]. (see Sect. 9.4 for a more detailed overview of data)

Test level	Relative effort
Module test	16%
Functional test	14%
Integration test	13%
Acceptance test	9%
Total	**52%** (incl. unit test) **36%** (excl. unit test)

TMap/TMapNext provide a relative distribution of the phases of a test project. Table 9.3 shows a wide distribution, which shows how difficult it is to get good generic metrics.

Table 9.3 A relative distribution of the phases of a test project. Source: [Koomen et al, 2007], [Pol et al, 1999]

Phase	Relative effort	TestGoal step
Preparation	6% -21%	± 1 & 2
Specification	33% -54%	± 3 & 4
Execution	21% -45%	± 5
Completion	2%-5%	± 6
Planning and control	15%-17%	in all steps

It is preferable to combine two types of controls to validate the estimate. For example:

Peer Review or Independent Estimate
It is easier and faster to control the estimate by means of a peer review than by having an independent person create a new estimate. This method works well if the estimate is more or less trustworthy.
The advantage of an independent estimate is that the "sparring partner" is not influenced. This method can be used to increase the confidence in the budget.

In combination with

Experience Data or Statistics

It is always preferable to use experience data to do a validation, but the data is often not available. A good alternative is to use statistics from literature. The statistics provide a good idea of the reasonableness of the estimate, and provide good independent arguments to support the planning.

9.2.3   Assessing the Requisites

In addition to a budget for the execution of the work, additional things may also be required, such as components that are needed for the test environment. Depending on the situation, these costs can be fixed or variable, or possibly even non-existent because the organization provides the requisites without charging the test project.

The checklist in Table 9.4 shows a number of possible requisites. The composition of the test environment will differ depending on the test levels that are run. More information about this can be found in Chap. 15 Test Environment. Things can be added to the checklist if the situation so requires.

Table 9.4 Assessment requisites

Component	Subcomponent
Test environment	
Hardware	Server PCs Printer …
Software	Operating system Test tools Databases Word processing package Planning package Back-up / Restore Communication software …
Infrastructure	
Workplace	Rooms Meeting room Furniture (chairs, desks, tables, filing cabinets) Photocopiers Telephone
Network	Router Hub Bridge Gateway UTP cables Coax cable …

9.2.4   Establishing the Budget

The budget is established by adding up all of the defined costs, i.e the costs estimated for the activities and the requisites. Make sure the budget distinguishes between fixed and variable costs. This makes it easy to establish which costs a change in the project runtime will incur. This is explained in the below example:

Example 9.1

The following project costs have been defined:

Hourly costs: €450/hour
        (on average, the five employees book 160 hours a week)

Hardware: €5,000 one-off
License tool: €100 per user per month
Rent tool: €900 every two weeks

If the test project is extended by three weeks, the costs for the test project will increase by €218,300:

Extension	1 week	2 weeks	3 weeks	4 weeks	5 weeks
Hours	160 x €450	160 x €450	160 x €450	160 x €450	160 x €450
License	5 x €100				5 x €100
Rent	€900		€900		€900
Cumulative costs	€73,400	€145,400	€218,300	€290,300	€363,700

9.3   Test Planning

9.3.1   *Generic Planning*

In addition to the costs, the runtime is an important component of the anticipated goal. In the Goal description, we established what the available budget and the expected end date is. The customer will want to know whether the expected end date will be reached within budget. This will, of course, depend on the total number of hours estimated and on the availability of the resources.

Experience shows that a planning frequently goes through several iterations before it is approved. Several versions are often needed before the customer approves it. Because a lot of effort goes into creating a detailed planning, a generic planning is created first. This planning is easily changed because it does not take the following items into account:

- Resource availability
- Dependencies between the subactivities
- Parallel projects

9.3.2   *Detailed Planning*

A detailed planning is developed after the budget and the generic planning have been approved. When the planning is created, the activities are assigned to employees. Their actual availability is investigated and it is established when the activities can be carried out.

The planning takes milestones, moments of decision and risks into account. Moments of decision and milestones often fall together at the end of a phase, but they don't have to.

Moments of Decision
Moments in which a choice is made, for example, a pilot that is run to gain experience with a new test tool. After a week, the pilot is evaluated. This is the actual moment of decision. If the new test tool meets the requirements and is easy to use, it will be used during the rest of the test project; if not, the current test tool will be used.

Milestones
A milestone is a point at which the progress of the test project can be clearly established. A milestone is often planned at the end of a phase, the delivery of a product or a moment of decision. An example of a milestone is a detailed test plan that has been approved or a smoke test that has been completed with a positive advice.

Project Risks
In contrast to the product risks, which are part of the test risk analysis, project risks do not say anything about the test object. Project risks indicate where the progress or the success of the project is in jeopardy. When describing the risks, we specify:

- What the risk is
- How the risk can be recognized if it occurs
- The impact of the risk
- The measure to be taken

In principle, planning a test project is not much different than planning a software development project.

A lot of literature is available on project planning. The following remarks, which are specific to a test project, build on the literature.

External Dependencies

A test project is strongly dependent on the development project. The quality of the test design depends on the stability and the quality of the test base.

If the test base changes frequently, time will have to be allocated to adapt the test design. The test's runtime depends on the availability and stability of the test object and the test environment. Experience shows that if the development project's build phase takes too long, the test object will not be ready for testing as planned. Moreover, the stability is often such that problems would occur were the team to continue working. Keep a close eye on what happens in the development project and indicate what the effects will be if a previous activity overruns.

Quality of the Software

The quality of the software impacts the number of test hours required and the runtime. Each error has to be recorded, discussed and retested. This takes time. Between recording and retesting, time is needed to solve the error and make the code available again. For module tests, the developer commonly executes the tests, meaning he can fix errors himself. This is not common for all other test types, and makes error fixing an external dependency.

This has to be taken into account when creating the planning. For example, estimate the expected quality of the system. If the quality differs from expectation, inform the customer. The risk that the planning is not met is suddenly becoming real. You have to look for a solution with the customer.

The following signs indicate that the quality of the software may be lower than expected:

- The previous test levels were not run or were not run as thoroughly as initially planned.
- The sanity check shows that the test base is incomplete or ambiguous.
- The smoke test reveals a lot of problems.
- The first day of the test produces a high number of errors.
- A new version of the software is needed at the beginning of the test.

Available Time

In general, tests are run at the end of the development project, Experience shows that preceding activities are often overrun. The go-live date is generally not delayed because business interests are involved, meaning that any additional time an activity takes will reduce the available test time. This is why it is important that the most important tests, which are defined in the risk analysis, are run at the beginning of the test phase.

Changing Requirements

The demands the customer makes on the test object often change during the development project. This can be the case because the market continues to develop, because of a growing insight into the way the system works, or because new technologies emerge. When the requirements change, it is almost certain that the plans and the test base will have to be adapted (also see the next item: "System changes"). This can mean that the system will go live with less functionality than planned. It is important to know this on time to prevent valuable time being wasted on testing functions that will not be included in the live product.

System Changes

The system and the test base go through many changes during development. The development time is estimated when change requests are planned. The development team is often granted additional time to implement the changes. Remember that a change also requires additional testing time and make sure that this time is included in the planning. This is not something that comes naturally to every organization.

External Parties

When testing with external parties, remember that they have their own planning and runtime. For chain tests, for example, all of the parties in the chain have to be ready on time. If the planning is too tight, the chances are high that the entire chain test will be overrun because one party in the chain is not ready on the agreed date.

Effective Productivity

Use effective productivity to convert the hour estimate to the runtime. An employee is never one hundred percent productive: they attend department meetings, take coffee breaks, go on vacation, etc.

Depending on the type of organization, the effective productivity is between 60 and 80 percent [DeMarco, 1999]. This means that an activity that takes one person seven hours to complete is not finished in an eight-hour workday.

9.4   Key Indicators

To create a good hour estimate, it is important to check the correlation between the planned activities. For the latter, Table 9.5 includes the necessary key indicators. The table is taken from the benchmark study of a few hundred IT projects in [Jones, 2000]. The percentages have to be used with care, but in practice they provide a useful crutch. A few remarks:

Table 9.5 The correlation between planned activities for a good hour estimate

Phase	Activity	Jones	Own Organization
Architecture	Architectural sanity check	2%	
		= 2%	
Analysis	Functional specifications	4%	
	Demo validation		
		= 4%	
Design	Initial system design	3%	
	Detailed system design	4%	
	Review		
	Reuse acquisition	1%	
		= 8%	
Realization	Building (coding)	17%	
	Module tests	**16%**	
		= 33%	
Tests	**Functional tests**	**14%**	
	Integration tests	**13%**	
		= 27%	
Acceptance	**Acceptance tests**	**9%**	
		= 9%	
Deployment	Integration		
	Configuration management	2%	
		= 2%	
General	Documentation	4%	
	Project management	11%	
		= 15%	
		= 100%	

- The percentages provide an average over a large number of very different projects.
- The table is based on the creation of management information systems. For other areas, other key indicators may apply.
- New development methods can produce other key indicators.
- Feasibility studies and pilot phases are not included in the total.

Many organizations have their own experience data. Although this data is the most interesting as benchmark for the budget, differences between the projects need to be taken into account here too.

10 Test Plan

10.1 Introduction

Testing is usually done on a project basis. In order to steer toward the goal, an approach is also created for testing: the test plan. The test project is seen as an independent project or as a subproject of a development project, whereby we assume that all of the test activities are carried out in the test project. The test plan can then be considered as a subplan within the framework of the bigger project that is specifically aimed at testing. We distinguish between an MTP and a DTP.

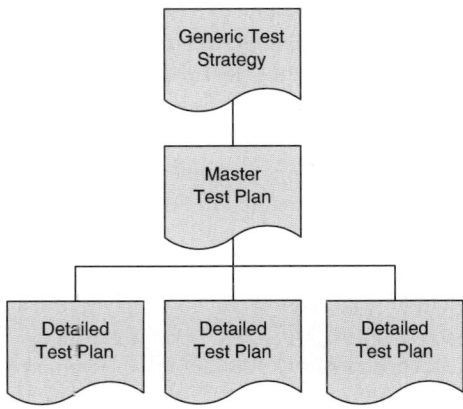

A master test plan is made on a more strategic level and deals with more than one test level. It converts the generic test strategy into a test approach for all test projects that are within the scope of the project. The master test plan is generally created by the test manager.

A detailed test plan focuses on one test level. In principle, it has the same structure as the master test plan, but it deals with one specific test

D.-J. de Grood, *TestGoal*,
DOI: 10.1007/978-3-540-78828-7_10, © Collis B.V., Leiden, The Netherlands, 2008

level and with how it is carried out. The detailed test plan is usually created by the test coordinator.

Example 10.1: MTP and DTP or Connecta

The test levels used in the Connecta project are described in Example 4.1. A module test, system test, chain test and a user acceptance test are run on the project. Yasmin Hassouni, the test manager, is responsible for the execution of these test levels. She has appointed a test coordinator for each test level. We met David Bloom in Example 6.2. He will coordinate the system test.

The test manager creates an MTP, which provides an overview of the test levels that are run on the Connecta project. It defines a number of generic things for the project, including the tools that are used, how errors are recorded and processed, and which test environments are available. The MTP also shows what is expected of the system test, the test level for which David is responsible.

David creates a DTP for the system tests. He uses the framework established in the MTP to fill in the details of the system test.

This chapter discusses the creation of a detailed test plan. A number of templates are available, many of which primarily show which topics should be included in the plan. This chapter not only describes what should be included in a test plan, but also explains why certain topics are included. Experience shows that the instructions provided in this chapter enable even a less experienced tester to create a good test plan. Example 10.2 displays the index of an extensive test plan. Each of the elements will be explained in this chapter.

Example 10.2: Index of a detailed test plan

Task description
Test base
Test strategy
     Description of the test approach
     Test risk analysis
     Quality attributes
     Strategy matrix
     Technique matrix

Previous and next phases
Test environment
Quality assurance
Release advice
Change and error management
Transfer
Planning
Test organization
Organization chart
Responsibilities
Meeting structures
Products
Requisites
Changes and deviations

10.2   Description of the Assignment

The Goal description was defined during the Goal assessment (step 1 in the TestGoal step plan). The Goal description describes *what* the anticipated goal is; the Test plan explains *how* the anticipated goal will be achieved. The overall assignment description is therefore part of the test plan.

The test object is identified in the Goal description. Additional clarity is provided by indicating which systems will and will not be tested in a context diagram. Figure 10.1 displays a sample system architecture. In

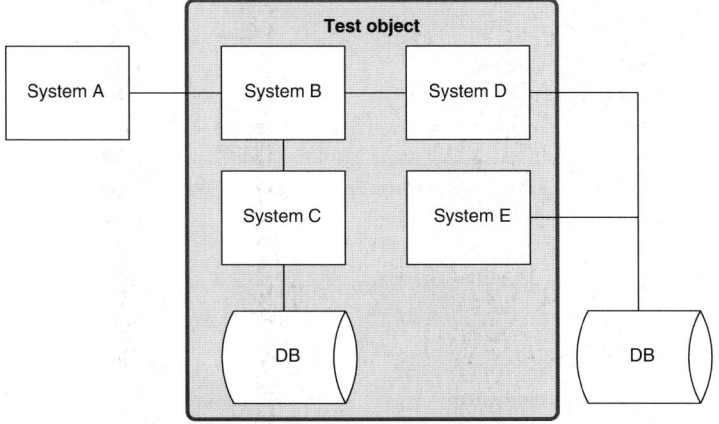

Fig. 10.1 A sample system architecture

this figure, we focus on the test object, also known as the system under test (SUT).

The Master Test Plan

In contrast to the detailed test plan, the master test plan only outlines how testing will be done. It does, however, define the test phases and the sequence in which they are run. The master test plan also has to indicate in a table or in a graphic which test levels will be run on which systems. In Fig. 10.2 the following activities are carried out:

- Functional acceptance tests (FAT) on systems B, C and D
- User acceptance test (UAT) on systems C and E
- Chain test (CT) on systems B, C, D and E

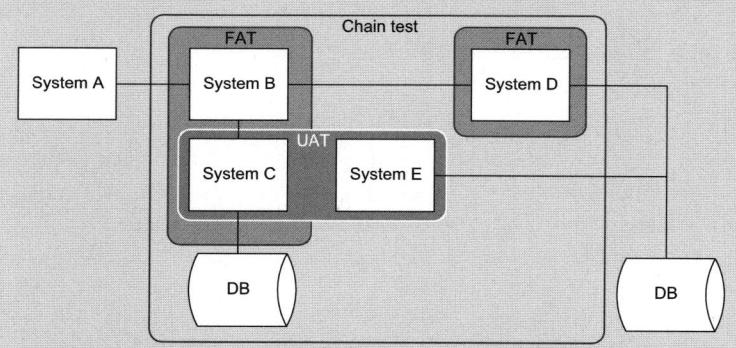

Fig. 10.2 A master test plan indicates which test levels are carried out on which systems

10.3   Test Base

The test base consists of all of the test object descriptions that are used as the basis for the tests.

For new projects on which tests have not yet been run, the test base consists of documentation, namely the system specifications. These are generally available in the form of

Requirement specifications

- A functional design
- A technical design
- UML diagrams
- A database design
- Interface specifications
- Important quality attributes
- Acceptance criteria
- Use cases and user scenarios
- Possible norms, standards, policies and legislation

Not every document is used in each test level. A module test will test technical specifications like UML diagrams, and the user acceptance test will generally cover user scenarios. This doesn't make the other documents any less interesting. If the tester wants to be a good interlocutor for the organization's various stakeholders he will have to understand the system both technically and functionally. He should therefore be familiar with both the high-level documentation and the technical documentation.

When a new system is built to replace an old system, the old system is sometimes used as a reference and is also considered part of the test base. If tests have already been run, it is likely that reusable testware is available. The testware's reusability is determined during the sanity check. If the testware is not usable without modification, the time to make the changes is included in the planning. The test base is included in the detailed test plan because the hour estimate, the planning, the test risk analysis and the tests to be designed are based on it.

Changes to the Test Base
In most projects, the system specifications change gradually. When this happens, the impact the changes have on testing need to be examined. If the impact is high, the initial planning becomes invalid. The test base that is included in the test plan serves as a reference to the versions that are the foundation of the initial test approach and planning. This reference is useful if we need to illustrate why the current test approach and planning differ from the original plan.

The contents of the test base should be controlled from the sanity check onwards. A good change management procedure that lays down how we deal with changes to the test base (see Sect. 10.4.10) is indispensable. The accompanying availability and quality are important from the time the test base is used. The quality and stability of the test base directly influence the quality of the tests to be designed. If the quality is poor, a lot of errors and ambiguities will surface during the sanity check and test design phase that will delay the test design, make it less accurate, or result in changes. Changes can, however, also be a result of increasing insight or business developments.

10.4   Test Strategy

The test strategy describes how the anticipated goal is translated into the chosen test method. The test strategy is created by combining the organization's guidelines, the test risk analysis and the planning into a consistent whole. Knowledge of and experience in testing is needed to create a test strategy. It is specialist work that requires making choices and explaining them. This chapter deals with the components of the test strategy.

It is convenient if the organization already has a generic test strategy or a master test plan. If either describes the above-mentioned components, the test approach in the detailed test plan can be short and only needs to indicate where the chosen approach differs from the normal workflow. If the test strategy or master plan is either non-existent or incomplete, the detailed test plan will have to describe the strategy in more detail.

In many test methodologies, the test strategy primarily consists of

- distributing the areas of attention, the risks and the functions over the test types
- determining the thoroughness with which the areas of attention, the risks and the functions need to be tested [Koomen et al, 2007], [Pinkster et al, 2004].

Although this is correct, the test strategy quickly turns into a technical account of strategy matrices, risk categories, test design techniques and cross tables. The above approach overlooks a number of very important aspects.

It should be possible to explain the test strategy to the organization's stakeholders. The test strategy enables them to gain insight into the way in which the testers contribute to the anticipated goal. In the strategy, the tester explains which activities are undertaken to make this contribution and how he provides information about the extent to which the goal has been achieved. Testing is a *profession* and not all stakeholders are familiar with things like strategy matrices, test design techniques, quality attributes, test types and risk analyses.

A test strategy should not only show how the test cases are going to be developed, but also how the test process ties in with the software development method and process. The test strategy also describes how testing ties in with the rest of the organization and the applied processes.

A test strategy also has to be developed in organizations where testing is less mature. Even organizations that do not do risk-based testing and do not use test design techniques still have a need for a test strategy.

To give the stakeholders insight into the test strategy, the strategy should be short and clear to ensure that non-testers also understand what's going on and feel they are part of the testing process. An accessible test strategy builds bridges and confidence and is therefore goal driven.

10.4.1 Description of the Test Strategy

A detailed test strategy can be very extensive. Nevertheless, it is important to be able to explain the essence of the strategy to non-testers, like business management, who may not know anything about testing but would like to understand the approach. The test strategy should be unambiguous and no longer than one page [Quentin, 2006] so that the readers don't have to evaluate tables with percentages and figures but can focus on what the test process actually consists of [Hul, 2006]. Among other things, the strategy explains

- Which goal is pursued
- Which tests are run to ensure the goal is achieved
- How the risks that jeopardize the goal are dealt with
- How the link is made to the software development process and the development method
- Which advantages the chosen strategy has
- How the quality of the process is assured

The anticipated goal was examined in the first step of the step plan. The test strategy explains which tests are run to achieve the goal. When mapping out the risks, it should be explained how the results of the test risk analysis serve as a starting point for reviews, sanity checks and the test design to be developed. Finally, it explains which quality attributes are important and how they will be tested.

The development method will have a lot of influence on the test strategy. If the traditional waterfall model is used, it is reasonable to assume

that extensive system designs have been created. They are usually reasonably stable and can be used as input for the test design, enabling a detailed verification. With agile development techniques, the system design is less detailed and often subject to change, requiring a more flexible test strategy. On the one hand, the strategy explains how the changes should be dealt with, and on the other hand how a meaningful statement can be made about the quality even though detailed specifications are missing. Nowadays, systems are often developed incrementally and the system delivered piece by piece. In such cases, it is important to know how the increments are built and whether the tests that cover the most important risks can be run on the first increment.

> *Example 10.3: Incremental development*
>
> The system is developed incrementally. One or more business processes are delivered per increment. The test strategy ties in with this and establishes per increment whether the delivered business processes meet the criteria. The business processes that were tested in earlier increments are also checked for regression.

Further, it should be established whether the increments can be properly tested. Experience shows that testability is often forgotten when the increments are defined. For example, in order to run the tests, functionality is needed that is provided by a later increment. The strategy takes the impact of the development method on the test process into account and explains which choices need to be made in order to respond accordingly.

Before the stakeholders agree to the test activities, they will need to trust the approach. This trust is gained by giving the stakeholders insight into the strategy and by pointing out at what points in time they are involved in the decision-making process. A bit of advertising doesn't hurt. Show that the strategy is well crafted and show what the advantages of the chosen strategy are.

> *Example 10.4: Advantages of a chosen test strategy*
>
> - The most important business processes are tested first.
> - The most important processes are released early in the project. This boosts the confidence business and the users have in the success of the project.

- Users can be trained on the released processes at an early stage. This enables the users to get used to the new components in the software and the processes.
- Regression is continually monitored. This prevents surprises arising at the end of the project and contributes to a growing confidence in the final solution.
- Business processes are first tested functionally and then checked by the users. This prevents the UAT tests leaving a negative impression on the user organization because of errors that should have been found during the ST.

Work that is done well and on time produces reliable information about the quality of the system. Explain in the strategy which measures have been included in the test project to ensure that the quality of the test project is high. If reviews are done and methodologies used, include a few words about the tester's expertise. This quality assurance enables the stakeholders to trust the test project.

This description can be expanded with the following components:

- The results of the test risk analysis
- The relative importance of the quality attributes
- The strategy matrix
- An overview of the test design techniques to be applied
- The previous and next test projects
- The requirements for the test environment
- Quality assurance
- How the release advice is reached
- How changes and errors are dealt with
- How the transfer of knowledge and products is organized at the end of the test project

Each of the above-mentioned components is explained in the following sections.

10.4.2   Test risk analysis

The results of the test risk analysis (see Chap. 7) are described in the test plan together with the date on which the risk analysis was done and by whom.

Step 2 – Approach

Example 10.5: TRA results

On <date> the following people did a risk analysis:
Name, job title
Name, job title
The following risks and their relative importance have been acknowledged:

Risk category	Risk area	Relative importance
Critical	Route calculation-Standard calculation	270
	Navigation-Enter destination	150
High	Route calculation-Find alternative	117
	Accuracy	99
	Navigation-Favorites	80
Medium	User friendliness	65
	Route calculation-Route type	63
	Extra-Traffic jam information	45
	Performance	45
	Navigation-Recent destinations	20
Low	Navigation-Home	15
	Extra-Weather forecast	9
	Settings-Audio	8
	Settings-Maps	6
	Settings-Standard	6

10.4.3   *Quality Attributes*

It is hard to define which requirements a high-quality system has to
meet. Not only because the answer to this question depends on the anticipated goal, but also because everyone has their own ideas about
quality. To provide a frame of reference, quality attributes were developed to make quality discussable and help the stakeholders indicate
which attributes they really find important. The stakeholders determine
the relative importance of the various quality attributes. This gives the
tester a clear picture of what his tests need to focus on. The prioritized
quality aspects can be displayed in Table 10.1 [Zeist et al, 1996],
[Bouman, 2004].

Table 10.1 Quality attributes

Quality attribute	Description	Relative importance H M L
Functionality • Suitability • Accuracy • Interoperability • Compliance • Security • Traceability	Extent to which the system displays functionally correct behavior, i. e. the presence of functions and their specified characteristics.	O O O O O O O O O O O O O O O O O O
Reliability • Maturity • Fault tolerance • Recoverability • Availability • Degradability	Degree to which the system continues operating under the specified conditions during a specified period.	O O O O O O O O O O O O O O O
Usability • Understandability • Learnability • Operability • Explicitness • Customizability • Attractiveness • Clarity • Helpfulness • User friendliness	Degree to which the system is suitable for use.	O O O O O O O O O O O O O O O O O O O O O O O O O O O
Efficiency • Time behavior • Resource behavior	Degree to which the system performs well. Expressed in transaction speed and used capacity at a specified load.	O O O O O O
Maintainability • Analyzability • Changeability • Stability • Testability • Manageability • Reusability	The ease with which the system can be changed or the effort that is needed to make certain changes.	O O O O O O O O O O O O O O O O O O
Transferability • Adaptability • Installability • Conformance • Replaceability	The ease with which the software can be transferred from one environment to another.	O O O O O O O O O O O O

Step 2 – Approach

Considering the large number of attributes, it is important to separate the wheat from the chaff and determine the attributes that are really important together with the stakeholders. The most important attributes are usually those that the stakeholders want to receive reports on.

There are two ways to implement the most important quality attributes in the test design:

- The quality attribute is integrated with the functional tests.
- The quality attribute is tested in a separate test cluster.

The choice depends on the nature and the importance of the quality attribute. If the quality attribute is very important, or if it is difficult to test, it is best to create a separate test cluster.

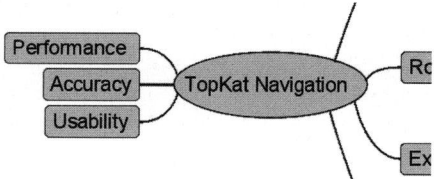

Fig. 10.3 Section of the test tree. On the left, the non-functional areas of attention have been included as separate test clusters.

A number of non-functional areas of attention have been included in the test tree that was created for the TRA (step 2 of the step plan). In Fig. 10.3, these are performance (time behavior), accuracy and user friendliness. Specialist knowledge is needed to test these areas of attention, which is why they are grouped separately. The test report provides insight into the test results, often per test cluster. The advantage of defining a separate test cluster for these quality aspects is that they appear as a recognizable unit in the test report (see Chap. 20 Test Report).

Quality attributes can also be integrated with the other tests if they are less important or if doing so is more efficient. In the example, the traceability test is integrated with the business process tests.

Example 10.6: Integration of quality attributes

An airline is building a Web site so it can sell tickets online. Traceability is considered to be an important aspect of selling tickets because the organization has a legal obligation to trace specific transactions, such as booked flights and payments.

In the test approach, the traceability test is integrated with the process flow tests. To test the traceability properly, a check has to be carried out to determine whether each situation has been recorded in the transaction log. To cover all situations, the business scenarios have to be run through in their entirety, as do the process flow tests. Integration saves a considerable amount of time.

For the process flow test, all of the business scenarios are run through. After each scenario, a check is carried out to determine whether the correct data can be traced to the transaction log. An additional test case has been added to the physical test design, after the test cases that describe the scenario. This test case describes how the check has to be carried out to determine whether the transactions and payments made in the scenario can be traced.

To illustrate which attributes are relevant at which point in the test project, a strategy matrix is created.

10.4.4   Strategy Matrix

In the strategy matrix, the relevant quality attributes are distributed over the test clusters or test levels. In the MTP, the quality attributes are distributed over the test levels. It is logical that the tester focus on user friendliness during the UAT, as it is not something he will pay much attention to during the module tests. A more interesting consideration can be made for the security and performance tests. Performance is often tested late in the development cycle, which is logical because that's when a working system that is representative of the live environment is available. Unfortunately, experience shows that performance often creates a problem that is discovered late. Suppose the performance of the system is a critical success factor. It is then a strategic choice to pay attention to performance during the module test, for example, by examining the efficiency of each database query. We can also verify early on how security is assured in the system design.

In the detailed test plan, the tester indicates how he will distribute the quality attributes, over the test clusters. The starting point is the quality attributes as they have been assigned to a test level in the master test plan. A login module, for example, is tested on security. The interface that offers large files that have to be processed by the system is tested on performance. If these files are supplied by various parties, it is wise to test if file processing meets the applicable standards. This conformity test is part of the quality attribute "transferability."

A matrix, such as the one displayed below, can be used to illustrate which attributes are relevant in which test cluster and at which test level [Hul, 2006].

Table 10.2 Sample strategy matrix. The matrix shows which risk area is tested for which quality attribute at which test level.

Risk area		Functionality	Accuracy	User friendliness	Performance	Interoperability	Clarity	...
		H	H	M	M	M	L	...
Route calculation-Standard calculation	C	ST (H)	MT (H) ST (M)	AT (M)	ST (M) AT (L)		AT (L)	
Navigation-Enter destination	C	ST (H)		AT (L)			AT (L)	
Route calculation-Find alternative	H	ST (M)			AT (M)			
Accuracy	H	-	-	-	-	-	-	
Navigation-Favorites	H	ST (M)	MT (H) ST (L)	AT (L)				
User friendliness	M	-	-	-	-	-	-	
Route calculation-Route type	M	ST (L)	MT (H) ST (L)		AT (L)			
Extra-Traffic jam information	M	ST (L)				MT (H) CT (M)		
...								

C=Critical, H=High, M=Medium, L=Low, N=Not relevant, MT=Module test, ST=System test, AT=Acceptance test, CT=Chain test

The two left columns indicate the priority of the risk areas. They are specified in the TRA (see Sect. 10.4.2). The two upper rows display the quality attributes with their relative importance. For combinations of quality attributes and risk areas, the matrix shows the test levels at which they will be tested. In the example, the importing of traffic jam information is tested on interoperability in the module test and the chain test. In the module test, the technical interface is thoroughly tested to demonstrate that all traffic jam information can be processed (priority is high). In the chain test, it is investigated if it is also possible to receive traffic jam information when it is transmitted by the interfacing system (priority is medium).

10.4.5 Technique Matrix

The test strategy also indicates for each test level and test cluster which test design techniques will be used. The test design techniques are discussed in Chap. 12 Logical Test Design. In the MTP, the techniques are grouped by test level. In the DTP, the techniques are grouped by test cluster or group of test clusters. The technique matrix is shown below.

Table 10.3 Sample technique matrix: The prescribed test design techniques are indicated for each test level.

Test level	Risk category Low	Risk category Medium	Risk category High	Risk category Critical
Module test	AT statement coverage	AT statement coverage	AT branch coverage	AT condition coverage
System test	EP valid Syntax	EP valid + invalid Syntax	BVA valid State Syntax	BVA valid + invalid Syntax State C/E
Acceptance test	Error guessing	Exploratory testing Load	Exploratory testing CRUD Load Stress	PCT branch coverage CRUD Load Stress Reliability Concurrency
Chain test	<not tested explicitly>	PCT statement	PCT branch	PCT branch

AT=Algorithm test, EP=Equivalence test, BVA=Boundary value analysis, PCT=Process cycle test

Read from left to right, Table 10.3 shows that the techniques get stronger. Components with a high risk are tested more thoroughly than components with a low risk. Read from top to bottom, the table shows that different test design techniques are used. At each test level, the test object is looked at from a different angle. This creates a better coverage because different things are tested at each test level.

10.4.6   *Previous and Next Phases*

A third part of the strategy is a clear phasing of the project. The V-model can be used to illustrate the phases. Specify what the previous and the next phases of the test project are, and then define the entry and exit criteria that apply to the project.

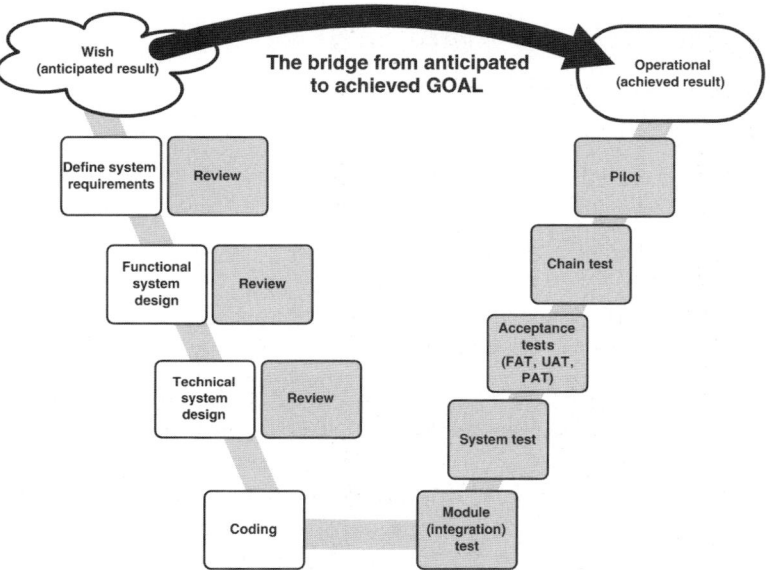

Fig. 10.4 Representation of the V-model. The V-model consists of two legs that, together, form the letter V. The left leg shows how the initial wish is converted into system requirements. The system requirements are converted into an increasingly detailed system design that serves as a basis for the programmer. The left leg also defines the static tests; each design is tested by means of reviews. The right leg shows the dynamic tests. The next test types, module (integration) test, system test, acceptance test, chain test and pilot can be recognized.

At the beginning of the test project, the output and the status of the previous phases will be important pieces of information. When test time is limited, everything should be done to ensure that the testers can do their jobs and rather than those that should have been done in a previous phase. This ensures that the little time that is available for the planned test activities is spent efficiently.

Identify the dependency the test project has on the previous activities and discuss important items with the responsible colleague. The DTP defines the things that must be taken care of in order to be able to start the test project without problems.

After a release advice, the test object moves on to the next phase, which is often the next test level or the live environment. Remember that the next party that will be working with the system will look at the testers in the same way that the testers in the previous phase looked at their predecessors. It is a good idea to determine when the test level is finished and what it means for the next phase (entry and exit criteria). In general, the test manager should have included this in his MTP. It doesn't hurt, however, if the test coordinator ensures that the process also ties in with the details.

Building bridges to the stakeholders of the other phases is crucial in order to ensure that the expectations are the same. It also prevents gaps or overlaps.

10.4.7 Test Environment

Different test projects put different demands on the test environment.

If it is important for the test project that errors are quickly fixed and re-tested, it's worth considering running the tests in the development environment. A development environment has a so-called fast deployment cycle. The time between finding an error and the availability of the solution is then short. If, however, similarity with the live environment is a deciding factor for the test project, an acceptance environment is better. Bear in mind, however, that it takes longer before solved errors can be retested in an acceptance environment because the deployment of the new software is more formal and hence takes more time. Because the acceptance environment resembles the live environment more closely, the likelihood that a component that works in the test environment will cause problems in the live environment is considerably reduced.

The requirements for the test environment are described in detail in the design step of the step plan (Chap. 15 Test Environment). In the test plan, these requirements are outlined insofar as they are part of the test strategy.

The maintenance of the test environment and the testware is importance to assure the quality of the test project. Include how this is done in the test plan, which parties are involved and how the maintenance is organized.

10.4.8   Assuring the Quality of the Test Project

The detailed test plan should also specify how the quality of the delivered products is assured. Think of the following measures and descriptions:

Review, Inspection and Walk-Through Sessions
The products of the test project will be delivered to stakeholders for review or inspection. The objective is twofold: on the one hand, the review or inspection should reduce the number of errors, and on the other hand, it should increase the involvement of the stakeholders. By acquiring knowledge of the contents of the products, they gain insight into the quality of the approach and the working method. For the latter objective, walk-through sessions can be done as well.

Provide feedback for each review comment in the next version of the document. Indicate how the comments have been processed, and if they haven't, explain why.

Sanity Check and Smoke Test
A sanity check and a smoke test are run. The purpose of these tests is to establish whether the objects are of sufficient quality and the planned test activity can be executed. The tests prevent time being lost during the execution of the test activities. If the objects are of lower quality than expected, the anticipated goal may be jeopardized. If the smoke test results in a negative advice, measures need to be taken. This limits the risks and ensures the planned activities can be carried out.

Testing the Testware
The testware will be tested for functionality and clarity. The resulting feedback will be processed where relevant.

Change Procedure
After the products of the test project have been approved, the change procedure is applied to ensure changes are carried out in a controlled manner.

Review of the Test Design
It is important that the stakeholders have an idea of the content of the tests so they can see for themselves that testing is done seriously, that their input is used in the test design, and that the points of attention that  are important to them will indeed be tested. Experience shows that stakeholders are more involved and more readily accept that testing takes time when they have insight into the content of the tests.

Moreover, knowledge of the tests that will be run helps the stakeholders evaluate the release advice. After all, how can someone have confidence in a release advice if they don't know how it is established?

Reviewing is a method that is frequently used to familiarize the stakeholders with the contents of the test project. One of the most extensive products of the test project is the physical test design. It's not easy to read either! It requires patience and knowledge of the system. Most users and managers would prefer to do anything rather than review a test design!
There are a number of ways to make the review more accessible to the stakeholders.

Divide the Reviews into Small Sessions
This minimizes the review effort. It is easier to start a review with a document of 20 pages than with a document of 200 pages. Experience shows that the last pages are usually reviewed less thoroughly than the first pages. Organizing a special review session for the last part of the document minimizes this effect.

Tell each reviewer which parts they should review
Only ask the reviewer to review specific parts. These are, of course, parts that fall within the reviewer's field and expertise. The advantage is that it takes the reviewer less time.

Use the Results from the Test Risk Analysis to Vary the Thoroughness of the Review
How thorough should the review of the test design be? This depends on the risk the test case covers. For less important test cases, it is very probable that the tester will only want to know if he is testing the right thing, in which case the review is only aimed at the test goals. For more important test cases, he will also want to know whether he is carrying out the test correctly: has he interpreted the specifications correctly, are the expected results accurate, is the test case going to work? By varying the thoroughness of the review, the reviewer can spend his valuable time on the right things.

Give Each Reviewer a Specific Scope
Let the reviewers look at the design from different perspectives. For example: the tester's colleagues look closely at how the tests have been developed and whether the test design techniques have been applied correctly. The analysts pay attention to errors in the contents of the test design and the users look at the coverage of functions that are important to them. This type of review is called "perspective-based reading."

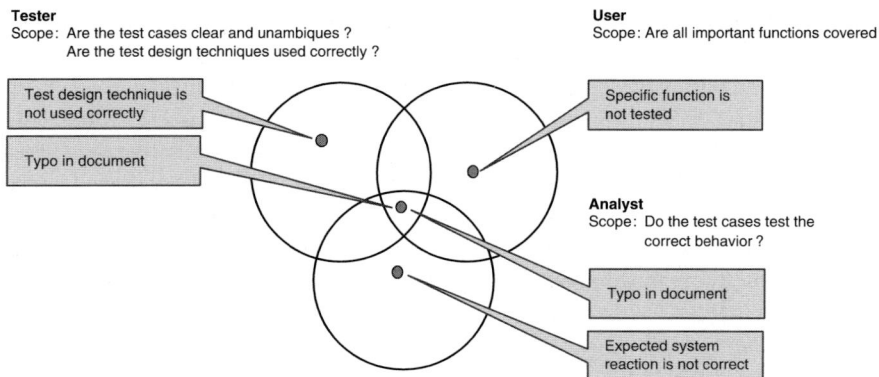

Fig. 10.5 The principle of "perspective-based reading." Each reviewer pays attention to things that are important to him. The advantage of this method is that the errors found during the review are more diverse. The above figure displays a number of errors. The typo could have been found by any reviewer. Other errors are always found by just one reviewer because he has focused on this particular issue [Basili]

This working method has the advantage of a clearly defined assignment and a clear definition of what is expected of the reviewer. It also increases the efficiency of the review because each reviewer can focus on his own field. It also increases the coverage of the review. The number of identical comments (for example, the typo in Fig. 10.5) that the review produces decreases.

Walk-Through Session Test Design

A walk-through session is a good way of giving others in the organization insight into the coverage of the tests. A walk-through session consists of two parts.

In the first part of the session, the test coordinator explains how the team approached the test design and how the stakeholders' input was incorporated. The test coordinator uses example to show how the TRA and test design techniques were applied. He demonstrates how the tests were elaborated and how the test results relate to the anticipated goal. The first part is finished as soon as the participants have a good understanding of the approach.

In the second part of the session, the participants get to ask questions. A frequently asked question is "Will you thoroughly test the function I find so important?" The test coordinator shows the extent to which the test will be run so the inquirer can establish whether he trusts the test design. Should the inquirer not trust the test design, it is further elaborated and an action item formulated. The test coordinator follows up on the action item and informs the participant about the result.

At the end of the walk-through session, the participants are asked whether they have confidence in the test design. Do they think that, when the defined tests have been run with a positive result, the system will be good enough to move on to the next phase? If the participants feel that there are still unaddressed risks, additional actions are discussed. An additional walk-through session may be needed after the action items have been dealt with. If the conclusion of the participants is positive and they trust the test approach and the tests, the test coordinator can be satisfied. The participants' confidence is important for the acceptance of the release advice.

> Make sure that the test manager attends the walk-through sessions as some of the participants' questions may be covered in other test levels. Because the test manager has an overview of all of the project's test projects he can confirm this on the spot or make sure another test coordinator is informed about the stakeholder's concern.

Step 2 – Approach

10.4.9 Release Advice

The test plan also contains information about the release advice that is given in the test report. A release advice can be negative, conditional or positive, and indicates whether the test object can move on to the next phase, which can be the next text level or the live environment.

The test plan explains how the release advice is established: by means of a formal test against fixed acceptance criteria or intuitively during a triage meeting. The most important stakeholders, which are specified in the Goal description, must attend the meeting.

Determine who ultimately gives the release advice. In a standard test project, this is usually the test coordinator or the test manager. Also agree on how the release advice will be treated. Thinking about subsequent actions in advance avoids tough discussions and panic if a negative release advice is given at the last minute. Describe what will happen if a negative release advice is given. Is the release advice binding, or is it, as is usually the case, an advice? Who decides to go live if the latter is the case?

10.4.10 Change and Error Management

Changes to the system, the test base and the testware have a lot of influence on the test project, which is why they need to be implemented in a controlled manner. Changes to the system and the test base also impact the software development project, which is why it is very likely that agreements on how to deal with changes will have already been made for the project. The test project will tie in with the process and system the organization uses. If a smooth process is not available, one should be defined and implemented.

It is not only important to know which system specifications have been used for the test design, it is also important that they have the same version as those the developers used to build the system. The following procedure has proven to work well: each change is implemented in the specifications by the analysts, who give the document a new version number and release it to the project leader. The project leader determines the moment at which the specifications will be used. He releases the specifications to the developers and the testers and determines for which software release the specifications are used as baseline. The baseline for developers is the test base for the tester.

The first advantage of this approach is that the analyst can implement his changes directly. Some changes turn out to be a lot more complex than they seemed to be at first sight. Dealing with the changes quickly means that the bottlenecks or unforeseen complications are detected and discussed at an early stage. The second advantage is that the specifications are released after having been verified, and that both the test team and the development team are working with the same versions.

Errors that are found during reviews, sanity checks, smoke tests and test runs are recorded. Once they've been recoded, it has to be determined whether they will be fixed and when. Include in the test plan how error management has been set up and who makes the decisions, or refer to the existing procedure (see also Chap. 19 Error Recording and Management).

10.4.11 Transfer

Before the test team is discharged, the testware and the acquired knowledge are transferred to the organization. Indicate in the test plan what is transferred to whom and what the rules are.

Example 10.7: Planning the transfer

During the system test on the Connecta project (also see Example 1.3 and further examples), products such as the test design and a review log were created. The testers also acquired knowledge about the new system. To ensure that the products and knowledge are reused as best as possible, they are transferred. During the approach phase, the test coordinator maps out which parties need a product or information and when. After consultation with the stakeholders, he includes the following in his DTP:

Test Design
Description
The test design is transferred to the test center. The test center will run future regression tests.
Recipient
Test center manager.
Transfer method
Cleaned test archive is transferred after an oral explanation.
Moment of transfer
In the Transfer step described in the step plan.

Review Errors in Test Base
Description
The issue list with the errors that were found in the test base is transferred to the team of analysts. The information can be used to improve the test base for future projects.
Recipient
Analyst team leader
Transfer method
The review log for the test base is regularly exchanged and discussed.
Moment of transfer
At fixed times throughout the test project.

System Configuration
Description
The knowledge of the system configuration is transferred to the maintenance organization. A lot of knowledge about how the system should be configured is gained while running the tests. Although the configuration was set up for testing purposes, the knowledge gained may also be applicable to the live environment.

Recipient
The system administrator, who is responsible for the configuration of the live system.
Transfer method
The system configuration is discussed with the system administrator.
Moment of transfer
In the Transfer step described in the step plan.

System Knowledge
Description
The test team has learned how to work with the system. The experience gained about the workings of the system and its peculiarities will be transferred to the subproject Implementation. The trainers can use the knowledge for a practical system introduction.
Recipient
Project leader/trainer of the subproject Implementation
Transfer method
Review and discuss the training before it is given.
Moment of transfer
When work is started on the training material.

Learning Points
Description
During the evaluation, the learning points are included in the learning points report.
Recipient
Test department manager, Test manager.
Transfer method
The lessons learned report is transferred. The recipient is present during the assessment.
Moment of transfer
During the last step of the project (assurance).

10.5   Planning

Chapter 9 explained the creation of the test budget and planning in detail. The approved test budget and planning are included in the test plan.

Some project management methods, like PRINCE2, store the planning in a separate document because it changes so often. If the planning is stored in a separate document, include the initial planning. Include a reference to the document in which the updated planning is stored.

10.6 Test Organization

The test plan contains an overview of the test set up. The test set up is included in the test plan to indicate which parties are involved in the test process and how communication between these parties is structured. It is important that all of the parties that are needed to achieve the defined milestones and to execute the planned activities are identified. So that everyone knows their place in the test project, the roles, responsibilities and tasks are determined for each party.

10.6.1 Organization Chart

Use organization charts to make things transparent. More complex projects may require more than one organization chart, for example, for the project organization and the line organization, and maybe an organization chart for the external customer or contractor.

The organization chart displays the hierarchical relationships and should contain a level above the customer. It is useful to know who is backing the customer in the event of escalation.

The organization chart is also useful to indicate communication lines. Communication lines may seem logical in a neatly arranged project, but when external parties are involved, special agreements may have been made. In politically sensitive situations, it is not always desirable that everyone exchange ideas freely. In such cases, define the topics the parties may and may not talk to each other about.

In Fig. 10.6, the dotted lines indicate that the testers communicate with the developers and that the test coordinator communicates with the external contractor's project leader.

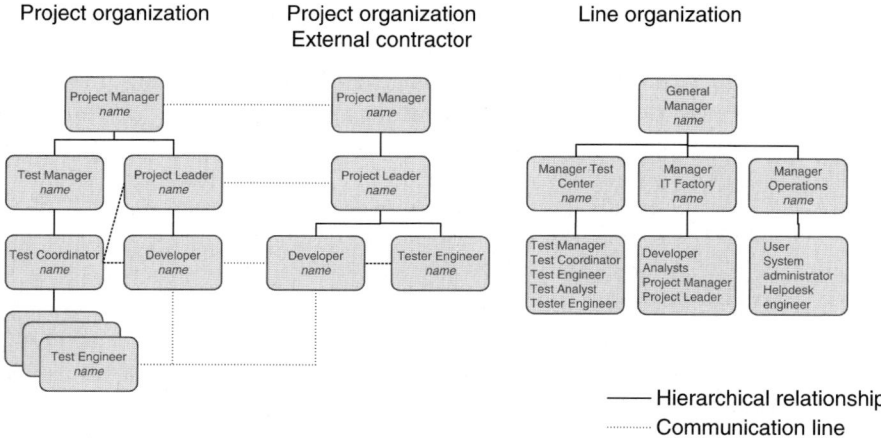

Fig. 10.6 An organization chart

10.6.2 Responsibilities

The description of the test organization also contains the description of the stakeholders' responsibilities, i. e. the employee's role in the project. State the person's job title and the responsibilities he can be approached about. Mapping out the responsibilities makes the interests transparent and helps understand why stakeholders react the way they do.

The RACI Chart

The test plan contains a description of the tasks and responsibilities. A RACI chart is a good way to display them. A RACI chart indicates which parties are responsible for the execution of which activities as well as who performs the action and who is informed or consulted.

When creating a RACI chart, activities and stakeholders are matched in a table. Codes indicate how each person is involved in the activity. The following codes are used to fill in the table:

- R = Responsible; performs the corresponding action.
- A = Accountable; person who is ultimately responsible for this activity.
- C = Consultation; is consulted during the execution of this activity.
- I = Informed; is informed about the result of this activity.

Part of a RACI chart is displayed in Table 10.4. The below RACI chart should be read as follows:

- The test coordinator is responsible for the development of the DTP (R). He does this in consultation with the testers in his team (C). If the test coordinator does not develop his plan correctly or completely, the test manager (A) will have a problem. The plan contains important information for the project manager, who receives the plan for information (I).
- The sanity check is run by the tester (R). The analyst has to deliver revised work. This also makes him responsible for the sanity check (A). The test coordinator is informed about the results (I).

Table 10.4 A RACI chart

	Project manager	Test manager	Test coordinator	Tester	Users	Maintenance organization	Analysts
MTP	A	R	C	I			
DTP	I	A	R	C			
Sanity check			I	R			A
Process review on test base	A						R
Configuration environment	A			I		R	
Test design			A	R	C		C
Test execution			A	R			
Test report	I		R	C			

10.6.3 Meeting Structures

The test organization needs structured meetings. It is useful to state which meetings will be held and to explain their objective so the available time can be used as efficiently as possible. When stating the names of the meetings, include the following items:

- Objective of the meeting
- Attendees
- Frequency
- Reporting
- Required time, including the time to create the report(s)

Test Progress Meeting
Objective: Establish the progress against the planning based
                on the test report
Attendees: Test coordinator, Project leader, Project leader
                external contractor
Frequency: Biweekly
Report: Test report
Time needed: No more than 1.5 hours per meeting

Standing Meeting
Objective: Establish progress, bottlenecks and dependencies
Attendees: Test coordinator, Test team members
Frequency: Every day during the test run
Report: None
Time needed: 15 minutes per meeting

It can, of course, happen that planned meetings are cancelled during the
test project or that additional meetings are needed. If regular meetings
with many stakeholders are cancelled, it may be desirable to adapt the
test plan. This does not apply to meetings that are held only a few times
or that have no strategic importance.

10.7   Deliverables

The products to be delivered by the test project are an essential part of
the test project's anticipated goal. In the planning phase, the customer
and the contractor discuss and agree on the deliverables. The agreements are described in the test plan.

A basis for mapping out the products is the TestGoal step plan. The
products from the step plan are included in the WBS checklist in
Chap. 9 Test Budget and Planning.

The product overview is the basis for the discharge of the people who
worked on the test project. The team can be discharged if the agreed result has been achieved; this means that all of the agreed products have
been delivered to the customer.

10.8 Requisites for the Test Process

Requisites are everything that is needed to be able to start, run and finish the test project, such as a test environment, workplaces or an additional printer. When requesting requisites, bear in mind that a number of items may have a delivery delay. In some organizations, requesting a mainframe test environment has a delivery delay of several months.

This is why requisites need to be planned. It is also wise to define their requirements. The requirements for a chair probably won't be all that exciting, but they will be for the test environment. This is why the requirements for the test environment are taken into account in the design phase (also see Chap. 15).

Step 2 – Approach

10.9 Changes and Deviations

Experience shows that a project's scope is subject to change. A wish to deviate from the developed and approved plan can suddenly arise. This is why the test plan describes how changes and deviations are dealt with. For example:

> Changes and deviations to the approach described in the test plan will be implemented in consultation with the customer and the contractor.

If need be, the test coordinator and the customer will work together to establish what the change is, how the stakeholders will be informed and how the change will be formalized. This can be done, for example, by creating a new version of the test plan.

Step 3 – Design

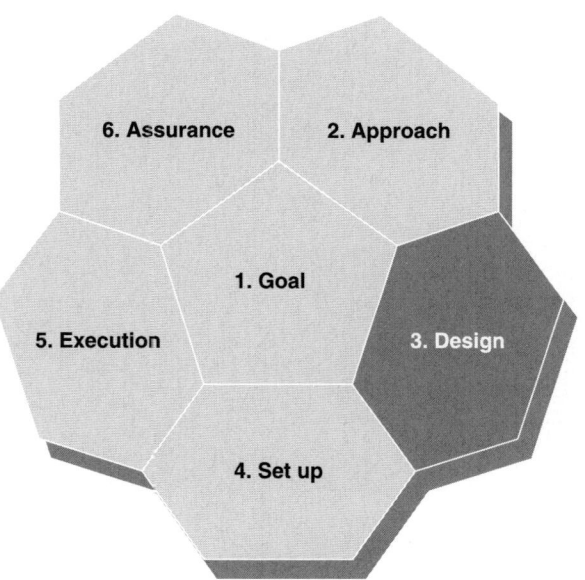

The Design step discusses the tests that need to be run and the environment that is required to run them. In this step, a logical and physical test design is developed, and the requirements for the test environment are established. The definition of the required test data is also part of this step.

The Design step consists of the following activities and products:

Activity	Product
Sanity check	Review log test base Sanity check report
Test designs	Logical test design Physical test design Test data
Requirements test environment	Requirements test environment

11   Sanity Check

11.1   Introduction

The sanity check provides insight into the quality and usability of the test base. It shows which measures need to be taken to start working on the test design without risks. The sanity check ensures that project risks are discovered at an early stage and that errors in the test base are dealt with. The sanity check is not intended to bring the project to a grinding halt, but to make risks discussable. The sanity check is run on both the test base and the testware. Reusing existing test designs is part of the TestGoal philosophy.

While the Goal description was created, the contents of the test base were also mapped out: relevant system designs, test designs and test tools. A first scan was also carried out to determine the test approach and the planning. The test base will be used actively during the creation of the test design. Before work on the test design can start, the last versions of the required documents must be available. It is on this version that the sanity check is performed.

The sanity check verifies whether the test base clearly describes the anticipated goal. This being the case, the test base can be used as a basis for the test cases. If testware is available, the extent of its reusability is checked. The testware must be of sufficient quality, complete, and up to date. The checklist is filled out during the sanity check. This ensures that key components are not forgotten. To fill out the checklist correctly, the test base needs to be thoroughly assessed, which is why this activity is combined with a review. Because the review is a risk-based activity, the TRA is used to determine which parts are more or less important. The depth of the review is adapted to the importance to ensure

D.-J. de Grood, *TestGoal*,
DOI: 10.1007/978-3-540-78828-7_11, © Collis B.V., Leiden, The Netherlands, 2008

that the important functions are reviewed more thoroughly than the less important ones. The sanity check produces

- Review findings
- A completed checklist containing:
  - A conclusion on the usability of the test base.
  - An overview of the measures that need to be taken so that the test design can be started without risks. If the test base does not meet the standards, the incomplete points are indicated. The risks for the test design and the planning, and the measures needed to cover the risks are indicated for each item.

The following sections discuss using the checklist and the review of the test base in more detail.

11.2   Filling out the Sanity Check Checklist

The checklist produced by the sanity check consists of two parts: one part for checking the design and one for checking the testware. The test analyst typically fills out the design checklist because he is the one who will be converting the system design into test cases. The testware

Table 11.1 Examples of errors, risks and measures

Error	Risk	Measure
System design is incomplete	Ambiguity about the desired result. High likelihood of changes in the specifications. High likelihood of errors in the system.	Map out the anticipated goal together with the stakeholders. Ensure that all of the team's test experts are informed.
System design contains numerous errors	High likelihood of changes in the specifications.	Explain that it is necessary to adapt the specifications before work can start on the test design.
System design is not up to date	High likelihood of errors in the system.	Explain that the project must take more releases and additional maintenance on the testware into account.
The software to be reused is not detailed enough	Tests do not give a good impression of the quality.	Plan an activity to adapt the testware.
The software to be reused contains numerous errors	Tests give an incorrect impression of the quality.	Increase the team's subject-matter knowledge so that the team can assess the system properly.
The software to be reused is not up to date		

checklist is typically filled out by the test engineer because he usually has knowledge of test tools and the physical test design.

On the one hand, the checklist is used to determine whether the test base contains all of the desired components, and on the other hand it is used to determine whether the components are of sufficient quality. All of the checks that are carried out are checked off in the checklist.

The test base may not contain all of the components. Components may be missing because they are not needed for the level of functionality being tested. An example is the checks run on reports: not every system generates a report. If the quality of the test is not good enough, the points that are of poor quality are indicated as well as the risks for the anticipated goal and how they jeopardize the efficient development of the test design. This is accompanied by a proposal for the measures that are required to cover the risks.

Test base checklist: Design
Conclusion

Conclusion	Y / N
The specifications (test base) are sufficiently detailed to start a structured test project.	X

If the conclusion is negative:
The overall findings of the sanity check are described below. The risks for the test design and the activiti that are required to cover them are specified for each point.

Findings	Risk	Measures
There are no performance requirements for the system.	Performance is specified in the TRA as an important quality attribute. It is not clear if performance requirements were taken into account.	*Analyst:* Define performance requirements so the performance can be accepted. *Analyst:* Explain to the test team how performance has been included in the design and where the performance risks are in the system.

Goal

Description	Y/N	N/A	Solutions
All of the functionality, processes and their coherence are sufficiently described.	√		
All of the functionality, processes and their coherence are consistent with the anticipated goal.	√		
All of the quality criteria have been sufficiently discussed.	X		Performance requirements have not been defined.
The results of the risk analysis have been traceably processed in the test base.	√		

Fig. 11.1 Part of a completed sanity check checklist

Step 3 – Design

The sanity check is finalized with a conclusion on the usability of the test base. A positive conclusion indicates that the examined specifications are a good and useful basis for the test design and test execution.

The completed checklist forms the sanity check report, which is used to determine who will take which measures and if work can start on the test design.

11.3   Continuous Learning

The sanity check checklist (see Appendixes A and B) can be supplemented and changed as needed. Experience from the test project and experience from previous sanity checks or other comparable projects can be processed in the checklist. Adding these experiences enables sanity checks to be carried out according to the latest insights and the acquired knowledge to be optimally reused.

11.4   Test Base Review

The contents of the system and test design are also reviewed during the sanity check. The system design review enables design errors and flaws to be detected and fixed early on in the project with relatively little effort and cost. The relative cost of fixing errors in the live product that are the result of specification flaws is much higher.

Experience shows that the review produces the most results if it is done when the design is almost finished. Most of the functionality will have been described, but changes can still be made to the design with little impact on the runtime. When 70 to 75 percent of the design is complete, the author will have had enough time to describe the functionality. He then knows how it has to be implemented and will have put the design on paper (perhaps after a couple of internal reviews with peers). At this point, the reviewer can get a clear picture of what the described functionality contains. See also Sect. 10.4.8, Assuring the quality of the test project, for practical tips on how to make the review more accessible.

Example 11.1: Summary of testability review

Evaluate for each component in the system design whether the component contains sufficient information to check whether the implementation meets the standards. Check whether the component is detailed enough and unambiguously formulated in order to derive the required test cases. Ask the following questions [Basili]:

- Is all of the information required to establish acceptance criteria for the requirement available? Is there enough information to define tests for each aspect of the requirement?
- Is there another requirement for which the same test could be defined, but of which the expected result is contradictory?
- Is it certain that the result of the test is easy to predict? Is the value of the output and the unit in which it will be represented known?
- Are there other interpretations that the programmer may use to build the system? Which test needs to be developed to demonstrate that the builder has indeed used a different interpretation? Is enough information available to define this test?

It is also useful to verify if the requirement is sensible from the view point of the anticipated goal.

- Does the requirement tie in with the description of the function? Does the requirement tie in with the system? Does the requirement contribute to the anticipated goal?

The test design review focuses on the testware that is already available. For example, the review establishes the extent to which the available tests are reusable. Often, the test team will have already worked with the test design, so its level of detail and quality are probably known. In such cases, the review can be limited to determining how up to date the design is and whether all of the system changes were implemented in the test design. Also bear in mind that the TRA may have changed. Pay particular attention to functions that were classified as "low risk" and are now classified as "high risk." It may very well be the case that a limited set of tests was defined for these functions, while a higher test depth is now required. It's also likely that these parts of the test design were reviewed less thoroughly in the past. If there is no knowledge of and experience using the test design, it will have to be reviewed thoroughly in order to prevent surprises arising during the test run.

Step 3 – Design

Errors are entered in the generic error log, if it exists. If it doesn't, a separate review log can be used. The completed review log is sent to the author and discussed with him if necessary. Where required, the author fixes the error in the system design. Experience shows that errors are often found because the reviewer has an incorrect picture of the system that is being built. In such cases, the author explains the error and answers the reviewer's questions.

11.5   Registration

The review errors are entered in the generic error log. If the log doesn't exist, a simple review error form can be used and sent together with the document to be reviewed. The below table displays the various elements that are on the review log form.

Table 11.2 Elements used in the review log

Element	Explanation
Number	Unique identification number of the error.
Document ID	Name, version number and author of the reviewed document.
Reference	Place in the document. Use agreed conventions, like Chapter 2, Sect 3, point 2.
Type	Type of error. The following categories are distinguished: Functional Error or inconsistency in the development Unclear Ambiguous description Incomplete Lacks functionality, or incomplete error handling Cosmetic For example, typos, terminology, version numbering, etc.
Description	Short and clear description of the error.
Prio	Priority of the error according to the effects on the product to be built if the statement remains in the specification. H = High – the statement definitely produces an error in the test. M = Medium – the statement may produce an error in the test. L = Low – the statement will not produce an error in the test.
Error status	Description of the status of the error. Open = New error, not dealt with Analysis = Needs to be analyzed On hold = Is temporarily set aside Decision = Waiting for a decision to be taken Action = Waiting for the result of an indicated action Finished = Has been processed
Proposed solution	The author describes the solution for the described error.
Change?	Here, the author indicates whether the statement will be dealt with (removing an error) or if the error has to go through error management (in the case of a change that falls outside the scope of the contract or a change that has great impact).

11.6   Formal Review and Inspection Procedures

If formal reviews and inspections are used in the organization, it is advisable to follow suit. Formal reviews and inspections are led by a trained moderator who supervises and guides the process. In addition, the moderator monitors the following things:

- Which review/inspection method is most relevant?
- Is the product ready for review or inspection?
- Is the maximum review speed applied (pages per hour)?
- Are the error logging meetings held according to the rules?
- Are the follow-up actions finished correctly?

If the organization is not yet familiar with formal reviews and inspections, it is advised to start with a less formal procedure (see also Sect. 10.4.8 Assuring the quality of the test project).

Step 3 – Design

12 Logical Test Design

12.1 Introduction

Step 3 – Design

This chapter discusses the development of a logical test design and the use of test design techniques. The logical test design is a collection of all of the logical test cases. Logical test cases define the tests on a logical level. The logical test design is based on the test base, the anticipated goal and the TRA. The aim is to create a balanced set of logical tests. "Balanced" means that the number of test cases that are defined for each function is in proportion to the complexity and the importance of the function. The distribution of test cases for each quality attribute is also carefully examined. Using the right test design techniques creates a set of logical tests that can be used to efficiently establish whether the system will contribute to the anticipated goal. The logical test design provides a balanced blueprint that is supported by the TRA. The blueprint can be used to elaborate the physical test cases.

> A logical test case describes *what* needs to be tested; a physical test case describes *how* it is done.

Why would you want to create a logical test design? A logical test design has a number of advantages. When developing the physical test cases directly from the test base, the tester has to divide his attention over two areas. On the one hand, he has to figure out which tests are necessary and on the other hand he has to describe the tests in detail. As described in Chap. 13 Physical Test Design, the physical test case describes step by step how the test has to be run. For example, the physical test case describes the screens that are tested and the test data that is

D.-J. de Grood, *TestGoal*,
DOI: 10.1007/978-3-540-78828-7_12, © Collis B.V., Leiden, The Netherlands, 2008

used. The development of a logical test case and the development of a physical test case are two different disciplines. The knowledge that is needed to develop logical test cases is different than the knowledge needed to develop physical test cases. If a logical test design is created in addition to a physical test design, the activities can be distributed over more people. Depending on experience and preference, a tester can deal with the more analytical logical test design, and someone else can concentrate on the more practical physical test design.

In addition, thinking about *what* without having to worry about *how* ensures that we do not apply unintended self-censorship. Although a test may be useful, it can be decided not to run it, perhaps because it is too difficult or because it too time consuming. In such cases, it is important to know that the test will not be run. Apparently the corresponding risk is not covered.

A logical test design contributes to reusable and maintainable testing. The logical test design provides insight into the tests to be run and distributes them over the test cases. To ensure the tests are run more efficiently, it can be useful to distribute tests for *one* logical test case over several physical test cases. Without a logical test design it is difficult to derive from the test cases whether the coverage is sufficient. The logical test design provides insight and makes it possible to explain the choice to include or exclude certain test cases in the physical design. This makes the test design maintainable and reusable.

Finally, creating logical tests enables the tester to vary the depth and focus of the tests so that the developed tests have maximum added value. The test can cover the identified risk by using test design techniques cleverly: only those tests are run that produce information about the suitability of the test object.

Is it always necessary to create a logical test design? The answer is, of course, no. A logical test design is only created if it adds value. It may not be important to know what the exact coverage of the less critical components is. A test case that is forgotten may not pose a problem and thus a logical test case is not needed. For the critical functions, however, more security is desired. The security for these components is provided by the logical test design.

There are a lot of organizations that use neither test design techniques nor a logical test design. In these organizations, the test base is often directly converted into a physical test design. But the tester will still implicitly use test design techniques during the conversion. Sound knowledge of the test design techniques and their principles still has added value.

Example 12.1: Implicit use of a test design technique

A tester is defining tests for an input field. He implicitly uses the test design technique BVA. In the test description, he defines the use of a number of boundary values. This increases the likelihood of finding errors and makes the test set more effective. The coverage of his test set is not entirely clear. Are all of the boundary values prescribed by the BVA design technique used in the test set or are only a few used? Are the boundary values that are and are not used traceable?

12.2   Test Design Techniques

This chapter describes the use of various test design techniques. Test design techniques are used to convert the test base into logical test cases.

A lot of different test design techniques are available. Applying all of the techniques at once is very time consuming and inefficient. A technique needs to be chosen. The suitability of a technique depends on a number of things:

- The type of error that can be expected
- The severity of the error if it occurs in the live system
- The information that is available to base the tests on

This is explained in the below figure.

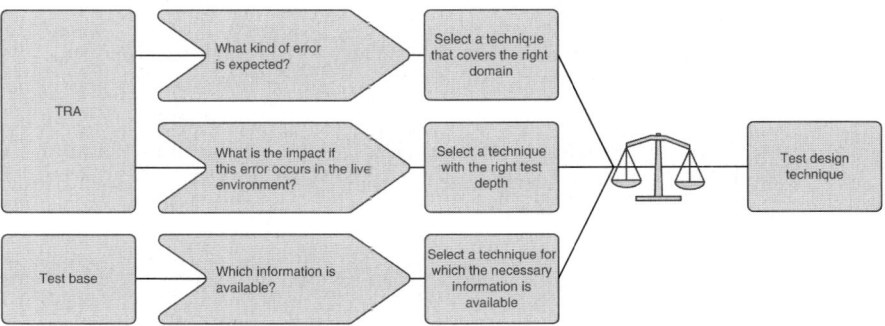

Fig. 12.1 Selection of test design techniques

Determining the Test Domain

Test design techniques force the tester to look at the test base and the test object in a specific way. Each test design technique has its own focus. Applying a specific technique increases the likelihood that errors of a certain type (within a specific domain) will be found. The structural approach of the technique ensures that the work is done more thoroughly and that errors are revealed that would otherwise remain unnoticed. However, the dark side of test design techniques lurks behind this advantage. Looking at the test object in a specific way makes you blind to errors that are outside the scope of the domain. These errors can be found with other techniques, which need to be chosen carefully and with the anticipated goal and the risks in mind.

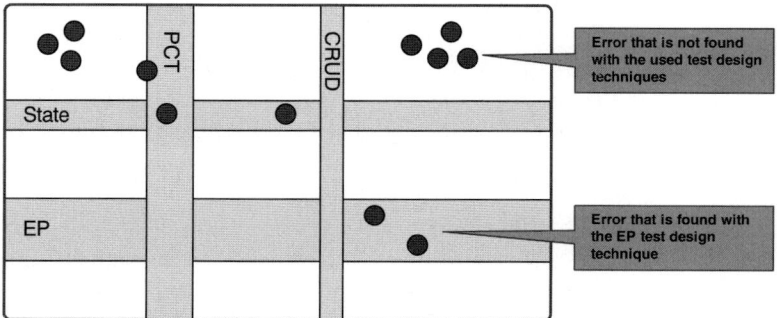

Fig. 12.2 Software errors in relation to the used test design techniques

In Fig. 12.2, the large rectangle represents the test object, which contains a number of errors. In this example, four test design techniques were used (PCT, CRUD, State Transition and EP) that reveal a number of these errors. In the example, the following things stand out:

- Errors that are found by one technique are not necessarily found by another technique.
- Some errors cannot be found with any of the chosen techniques.
- Errors often occur in clusters (for example, because of a poorly designed function, or because of a careless programmer).
- There are techniques that are very suitable in a specific situation (in this case EP) and techniques that are ineffective (in this case CRUD tests).

The TRA or the quality attributes can be used to determine the expected errors.

Example 12.2

A transaction processing system is being developed. In the system, certain actions are allowed or disallowed depending on the status of the transactions. The system also determines how long a transaction remains in a specific state. If necessary, the system displays a message to inform the user of the action he needs to perform.

The DTP of a test project includes the results of the TRA. One of the risks defined during the TRA was that the actions and messages may be triggered at the wrong moment.

In order to test this, all of the processes need to be run through and in each state checked whether the right messages are sent. The best way of doing this is by using state transition tests or process cycle tests. Both test design techniques mainly aim at state transitions and force the tester to go through the system in a structured manner.

The same DTP also indicates that reliability is an important quality attribute.

This is why a reliability test is suggested. This test design technique aims at demonstrating that the system can operate without problems over a long period of time.

Determining the Test Depth

Test design techniques make it possible to vary the test depth. Based on the available time and an estimate of the risks, it can be decided to test certain clusters more or less thoroughly. This variation is achieved by using heavy (strong) or light (weak) test design techniques. A technique is considered strong if it produces a lot of test cases. The advantage of a strong technique is that it has a wide coverage; the disadvantage is that the development and the test are more time consuming. The opposite can be said of a light technique. It results in relatively fewer test activities, but because the coverage is narrower, the likelihood that errors are overlooked is higher. Varying the test depth has the advantage that the test effort focuses on the most important components of the software. This prevents a lot of time being spent on testing less important clusters.

A stronger technique is commonly chosen for functions with a high risk than for functions with a low risk. Moreover, several test design techniques are often used for functions with a high risk.

Step 3 – Design

Example 12.3

The following two risks are clearly visible in Example 12.2:

- The wrong action can be carried out at the wrong time in the application.
- The system displays the wrong message. The message invites the user to perform an action he should not perform.

During the TRA, it was established that the impact of the first risk is greater than that of the second one. Therefore it is suggested to thoroughly test the triggering of the actions. A choice is made for a combination of state transition testing and process cycle testing techniques. The displaying of the messages has a lower risk and is therefore only tested with PCT.

Each test design technique has its own focus and aims at finding errors in a specific domain. A domain consists of errors of the same type. The test coverage of the cluster can be expanded by combining techniques so that different types of errors can be found. Depending on the test object and the expected errors, some domains will be more or less important. The test coverage of a cluster therefore depends on two variables: the strength of the test technique used and the coverage of the domain.

The technique matrix in the below table provides a starting point for the choice of the techniques. The domains are shown horizontally, the test strength is shown vertically. The table is based on best practices and is designed to show which techniques could be used for a cluster in

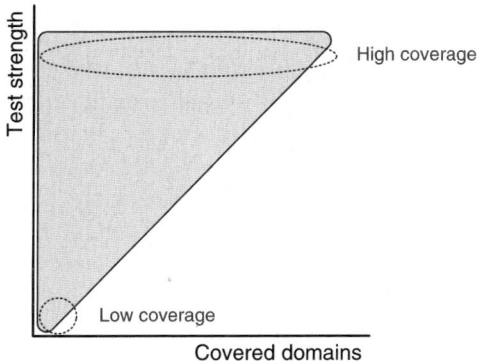

Fig. 12.3 The test coverage of a cluster depends on two variables: the strength of the technique used and the coverage of the domain

Table 12.1 Sample technique matrix: the domains are shown horizontally, the test strength is shown vertically.

Test strength	Input	Functional treatment				Data flow	Situation transitions			Processing time	Multi-user	Reliability		ET/HT
12	Syntax valid + invalid		BVA valid + invalid	C/E	AT condition coverage (or measure 2)	CRUD	PCT condition coverage	PCT measure 2	STT switch-coverage=1	Load	Concurrency	Stress	Reliability	ET/HT
10	Syntax valid + invalid		BVA valid + invalid	C/E	AT branch coverage/ measure 1	CRUD	PCT condition coverage	PCT measure 2	STT switch-coverage=0	Load	Concurrency	Stress	Reliability	ET/HT
8	Syntax valid + invalid		BVA valid + invalid		AT branch coverage/ measure 1	CRUD	PCT branch coverage / measure 1			Load	Concurrency	Stress	Reliability	ET/HT
6	Syntax valid + invalid	EP valid + invalid	BVA valid		AT statement coverage		PCT branch coverage / measure 1			Load	Concurrency	Stress		ET/HT
4	Syntax valid + invalid	EP valid					PCT branch coverage / measure 1			Load	Concurrency			ET/HT
2	Syntax valid						PCT statement coverage			Load	Concurrency			ET/HT

EP=Equivalence Partitioning, BVA=Boundary Value Analysis, C/E=Cause-effect Graphing, STT=State Transition Test, CRUD=Data Cycle Test, PCT=Process Cycle Test, AT=Algorithm Test, ET=Exploratory Test, HT=Heuristic Test

Step 3 – Design

a specific risk category. What is striking is that for a low test strength lighter test design techniques are suggested *and* that the testing covers fewer domains. For higher test strengths, the number of domains increases and a heavier (stronger) test technique is chosen. It is also worth noting that the coverage of a test technique can be changed by varying the test depth. The algorithm test has three test depths: statement, branch and condition coverage.

Determining Testability

Which of the techniques that are prescribed in the test strategy is used depends on the test base. A test technique may have a good focus and it may define tests with the right depth, but this does not guarantee that the system design contains the information that is needed to apply the technique.

Example 12.4

Syntax and state transition tests are suggested to test the application of a Web site where airline tickets can be purchased online.

Syntax tests check whether the restrictions on input data are implemented correctly. The test analyst checks whether the test base describes the criteria for the input fields and whether the input fields are mandatory. He discovers that the test base does not contain a data dictionary. The functional design describes a number of fields, but no restrictions have been defined for the majority of the input fields. It is not clear to the test analyst which restrictions he should test, and he doubts whether the syntax test is applicable at all.

The test coordinator suggests using the state transition technique to test navigating between screens. The system design clearly describes how a flight is selected and how a ticket is booked. A test analyst can use this description to derive how the screens succeed each another. But quick analysis reveals that a large number of alternative screen transitions are also possible. A state transition table (STD) is not available and the test base does not contain information about which transitions are allowed. The technique cannot be used without investing time to gather the missing information.

There are two options when the test base does not contain enough information:

- Find the information, no matter how hard it is
- Do not use the selected test design technique

The first option is preferred. Missing information or information that is hard to find is a risk. If one person cannot find the information, other people involved in the project will probably have the same problem. Who knows which system behavior will be implemented, who will decide whether the application is good enough? Ensuring that the information becomes available reduces the risk of surprises arising later on in the software development project.

But it's not always possible to obtain the missing information. There may not be enough time to adapt the system design, or the project manager may not think the risk is high enough to merit taking measures. In this case, the function will have to be tested in a different way. It is important that the risk is communicated clearly; after all, the tests will be less thorough than intended.

12.3 Use Test Design Techniques Cleverly

Using test design techniques has advantages, but should never become a goal in itself. Clever testing is more important than the formal use of a test design technique. When elaborating the physical test cases, a test can always be added that is not prescribed by the test design technique. This is all right as long as the test produces output. The same goes for eliminating test cases the technique prescribes but that are known to have little added value. Make sure the test case mentions why the test was omitted. This contributes to transparency and reusability. Moreover, the choice may be reconsidered during maintenance activities.

12.4 Little Experience with Test Design Techniques?

An organization that starts introducing test design techniques will probably not use all of the techniques described in this book. Our recommendation is to make small improvements first by applying simple test design techniques. A number of techniques are easy to learn and can usually be integrated with the existing strategy. For example, the BVA and syntax test techniques used to test input fields or incoming messages that need to be processed, or the PCT technique to test processes. The more complex and perhaps less obvious techniques can then be added to the test process at a later stage.

12.5   No Test Design Techniques

One reason for using test design techniques is to vary the test depth. The test risk analysis is used to determine which components are tested more or less thoroughly. By choosing a heavier test technique, a specific component can get more attention and a higher test coverage.

The test depth can still be varied, even without test design techniques. This is done by varying the process used to create the test design rather than the test design techniques. Below is an example from a generic test strategy, which shows how the approach ensures that the test effort is distributed evenly over the risks.

Example 12.5

In an organization, the following risk-based variation is applied during the development of the physical test design:

Risk category	Approach
Critical	For critical components, the test base is discussed by two test experts, a programmer, and an analyst or a user. During this discussion, the essence of the components function(s) is established. The group also checks whether there are undocumented exceptions that need to be included in the test. The minutes of the meeting provide a good overview of what should be tested and forms the basis for the logical test design. One tester develops the logical test design into a physical test design. Another tester carries out a review on the physical test design and discusses his findings with the designer. Experience shows that when two testers discuss a use case, half an hour is enough to split the main scenario up into one, two or more different test situations. This depth is often not achieved without discussion.
High	The component is discussed by two testers. The aim of the meeting is the same as for the critical components. The component is developed by one tester. If necessary, he gathers additional information from programmers, analysts or users. The other tester carries out a review on the physical test design and discusses his findings with the designer.
Medium	The component is developed by one tester without any prior discussions. The component is reviewed by a peer.
Low	The component is developed by one tester. Review is not compulsory.

12.6   Using Test Design Techniques

The previous sections discussed the how and why of a logical test design. It can be said that choosing good test design techniques helps define tests that will provide insight into the quality of the test object and help focus the test effort on the system components with the highest risk. Both are important, because the stakeholders want information on the extent to which the anticipated goal will be achieved as quickly as possible.

In the next section, a number of test design techniques are described. They were chosen according to their applicability. They are described according to a fixed format and contain examples to create an accessible overview of their use and to compare them. The explanation of the test design techniques is limited to the basic principle of the technique. There is, of course, a lot more to say about the techniques, but keeping the descriptions short prevents the focus shifting to details. For the less experienced tester, the description is a stepping stone to the use of test design techniques. For the experienced tester, the description is a reference, i. e. a quick reminder of how the technique should be applied.

The following techniques are described.

- Syntax testing
- EP: Equivalence Partitioning
- BVA: Boundary Value Analysis
- C/E: Cause-effect Graphing
- State Transition Testing
- CRUD Testing
- PCT: Process Cycle Test/AT: Algorithm Test
- Load Testing
- Stress Testing
- Reliability Testing
- Concurrency Testing
- HT: Heuristic Testing
- ET: Exploratory Testing

12.6.1   Syntax Testing

Description
A test design technique for a component or a system for which the test design is based on the syntax verification of the input. Or: syntax tests check

whether the limitations that are set for the input (and sometimes output as well) have been implemented correctly [ISEB practitioner, 2004].

Useful for

- Checking input fields on the user interface.
- Checking data elements in incoming and outgoing messages.
- Checking the consistency of the functional data model and the technical data model.

Approach

1. Determine the length and the restrictions for each data element (attribute) tested.
2. Determine the valid and invalid values for each attribute.
3. Design test cases according to these values.

Example

Step 1: Determine the attribute restrictions

A screen or a message contains three attributes: Name, Age and Children. The following restrictions apply:

Name   15 Characters, Optional
Age   3 Integer, Mandatory, maximum value 120
Children   Boolean, Mandatory

The functional specifications for a Web store for airline tickets contain the following:

- If the age is higher than or equal to 60, a 10% reduction is applied.
- Children up to 10 years get a 5% reduction.
- Frequent flyers get a free lunch.
- If the age is higher than 120, an error message is displayed.
- An error message is displayed when one or more attributes do not have the right size.

Step 2: Determine valid and invalid values

Valid and invalid values can be determined according to the applied character set and definitions for the various data elements. See the below example.

	Valid	Invalid
Name	Name ≤ 15 char Null	Name > 15 char Diacritical char
Age	0 ≤ age ≤ 120	1000 <0 Characters Null
Children	Y, N	Other Null

In this table, it is assumed that diacritical characters (characters with a mark placed above, below or through them such as é, ë, ç) are invalid, that the age cannot be negative or null and that the Boolean has the value Y or N. These assumptions need to be verified against the available specifications.

Step 3: Determine test cases

Combine the options determined in step 2 into a minimum number of cases that include each category at least once. A test case that has invalid values may have only one invalid attribute, otherwise it won't be clear which invalid value triggered the error message.
Example (the invalid values are in bold):

	Case 1	Case 2	Case 3	...	Case n
Name	Name ≤ 15 char	Name ≤ 15 char	**Name > 15 char**	...	Name ≤ 15 char
Age	0 ≤ Age ≤ 120	0 ≤ Age ≤ 120	0 ≤ Age ≤ 120	...	**Age = 1000**
Children	Y	N	Y	...	N

The above test cases are based on the three attributes and their restrictions. In addition, an error message is also displayed when one or more attributes are invalid. This information is used to define the result in the test case.

The table will then look as follows:

	Case 1	Case 2	Case 3	...	Case n
Name	Name ≤ 15 char	Name ≤ 15 char	**Name > 15 char**	...	Name ≤ 15 char
Age	0 ≤ Age ≤ 120	0 ≤ Age ≤ 120	0 ≤ Age ≤ 120	...	**Age = 1000**
Children	Y	N	Y	...	N
Result	No error	No error	Error	...	Error

Remarks

- *Test depth*: The number of invalid values that is included in the test is variable. The number can be infinite. A decision needs to be made about the extent to which the invalid values need to be tested.

> Use three syntactically invalid values for each attribute. Use more than three different values when you are not sure whether three values are enough, or when there is a strong likelihood that a lot of invalid values will be entered during use.

- *Difference with EP tests (see Sect. 12.6.2):* Children up to 10 years get a 5% reduction. When using EP, a distinction is made between Age < 10 and Age ≥ 10. A syntax test does not make this distinction: Age = 8 and Age = 43 produce different results (respectively a reduction of 5% and no reduction), but they are both valid syntax values. However, syntax tests also find errors that would not be found with EP, namely errors in the syntactical check on input fields.

12.6.2   EP: Equivalence Partitioning

Description

Division of the data input domain in equivalence partitions. Each partition leads to a test case. An equivalence partition is a collection of possible input data that leads to the same type of processing [ISEB practitioner, 2004].

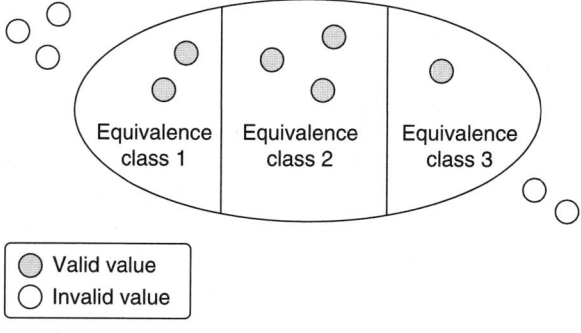

Useful for

- Checking input fields on the user interface.
- Checking functional decisions that are based on input data.

Approach
1. Determine attributes and relevant functional description.
2. Determine valid and invalid equivalence categories.
3. Determine test cases.

Example
Step 1: Determine attributes and relevant functional description

A screen or a message has three attributes, namely: Name, Age and Frequent flyer. The following restrictions apply:

Name   15 Characters, Optional
Age   3 Integer, Mandatory, Maximum value 120
Frequent flyer Boolean, Mandatory

The functional specifications of a Web store for flight tickets contain the following:

- If the age is higher than or equal to 60, a 10% reduction is applied.
- Children up to 10 years get a 5% reduction.
- Frequent flyers get a free lunch.
- If the age is higher than 120, an error message is displayed.
- An error message is displayed if one or more attributes do not have the right format.

Step 2: Determine valid and invalid equivalence categories

Use the system design to determine which input groups produce the same result. The starting point is the boundary values that are mentioned in the design. Using the above restrictions, the following boundary values apply for Age: 0, 10, 60, 120.

It is also defined which equivalence categories are valid and which are invalid. The following applies to each definition: the specified values are valid and the unspecified values are invalid. In the above example, an age of 130 is not an invalid value because the error handling for this input value has been specified. An age below 0 *is* an invalid value. The specifications do not provide an explanation.

The above section of specifications is used to create the following equivalence categories:

	Valid	Invalid
Name	Name ≤ 15 char Null	Name > 15 char
Age	0 ≤ Age < 10 10 ≤ Age < 60 60 ≤ Age ≤120 120 < Age	≤0 Null
Frequent flyer	Y, N	Null

Step 3: Determine test cases

Combine the options into a minimum number of test cases as described in step 2. For example (not all test cases are shown):

	Case 1	Case 2	Case 3	Case 4	...	Case n
Name	Name ≤ 15 char	Null	Name ≤ 15 char	Name ≤ 15 char	...	**Name > 15 char**
Age	0 ≤ Age < 10	10 ≤ Age < 60	60 ≤ Age ≤120	120 < Age	...	0 ≤ Age < 10
Frequent flyer	N	Y	N	Y	...	Y
Result	5% reduction No Lunch	No reduction Free lunch	10% reduction No Lunch	Error	...	Error

During testing, choose one input value from each equivalence partition. Normally, a tester chooses a value that lies approximately in the middle of the equivalence partition, for example, Age = 35 years. If the function is implemented according to the specifications, the choice for this value (within a partition) will not influence the result. It therefore does not matter whether 20, 35 or 43 years is entered.

Remarks

- *Difference with syntax testing:* In EP, a distinction is made between Age < 10 and Age ≥ 10. The syntax test does not make this distinction, Age = 8 and Age = 43 produce different results (respectively a 5% reduction and no reduction), but are both valid syntax values. Therefore EP is a stronger technique than syntax testing.

- *Difference with syntax testing 2:* Both EP and syntax testing distinguish between valid and invalid input values. The scope of both test design techniques is, however, different. Syntax tests mainly serve to test input: use of incorrect values such as diacritical characters, use of characters where numeric input is expected, or incorrect date formats. EP mainly serves to look at the functional decision that is based on the input. The emphasis is on entering the correct values. EP is used for syntactically correct values that are defined as invalid because the functional behavior has not been specified.

The processing of empty input (Null) is a different story. For mandatory fields, it can be interpreted as syntax verification. For optional fields, it is part of an equivalence partition because the functional processing does not change if the field remains empty. Depending on whether the number is mandatory, input "Null" is tested with syntax or EP testing.

- *Tip for making EP more effective:* The likelihood of finding a bug is increased if boundary values are used instead of the values from the middle of the equivalence partition, for example, 10 years instead of 43 years. This includes using elements from a different technique namely BVA (also see BVA testing).

- *Input and output EP:* The above example describes how EP can be used for input values. EP can, however, also be used for the output values. In practice, this means that each output is generated once as the result of choosing the right input values.

12.6.3 BVA: Boundary Value Analysis

Description
A design technique that consists of designing test cases that contain representative boundary values.

The approach is the same as that for Equivalence Partitioning, with the exception that the used values are not arbitrary values in the equivalence categories. With BVA, test values are chosen according to defined boundaries. These boundary values contain the values that come just below, on, and just above the boundary of the equivalence partition. A distinction is made between valid and invalid boundary values [Pol et al, 1999].

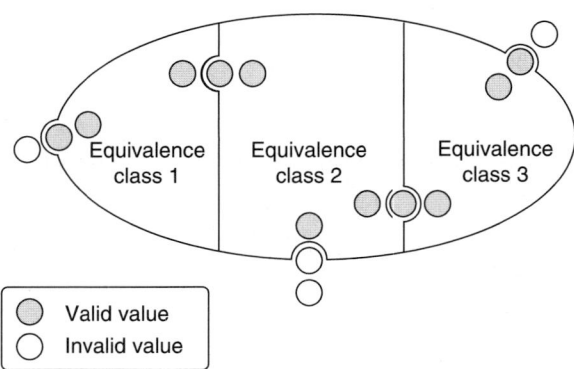

Useful for

- Checking input fields for user or message interface.
- Checking functional decisions made on data input.

Approach

1. Determine attributes and the relevant functional description.
2. Determine valid and invalid boundary values.
3. Define test cases.

Example
Step 1: Determine attributes and the relevant functional description

A screen or message has three attributes, namely: Name, Age and Frequent flyer. The following restrictions apply:

Name: 5 Characters, Mandatory
Age: 3 Numerical, Optional
Frequent flyer: Boolean, Mandatory

The functional specifications of a Web store for airline tickets contain the following:

- If the age is higher than or equal to 60, a 10% reduction is applied.
- Children up to 10 years get a 5% reduction.
- Frequent flyers get a free lunch.
- An error message is displayed if one or more attributes are invalid.

Step 2: Determine valid and invalid boundary values

Check which boundary values are defined in the design. Based on the above restrictions, the following boundary values are found for Age: 0,

10, 60, 999. With BVA, the boundary value and the values just below and above it are tested. For the boundary of 60 years, the values are 59, 60 and 61.

	Valid boundary value	Invalid boundary value
Name	1 char, 4 char, 5 char	6 char, 0 char
Age	0, 1, 9, 10, 11, 59, 60, 61, 999	-1, 1000, 1001
Frequent flyer	Y, N	

Step 3: Define test cases

Use the boundary value to combine the options determined in step 2 into a minimum number of test cases.

Valid boundary values:

	Case 1	Case 2	Case 3	Case 4	Case 5
Name	1 char	1 char	4 char	5 char	1 char
Age	0	1	9	10	11
Freq. flyer	N	Y	N	Y	Y
Result	5%	5%	5%	No reduction	No reduction
	No free lunch	Free lunch	No free lunch	Free lunch	Free lunch

	Case 6	Case 7	Case 8	Case 9
Name	4 char	5 char	4 char	5 char
Age	59	60	61	999
Freq. flyer	N	N	N	Y
Result	No reduction	10%	10%	10%
	No free lunch	No free lunch	No free lunch	Free lunch

Invalid boundary values (the invalid values are bold)

	Case 10	Case 11	Case 12	Case 13	Case 14
Name	**0 char**	**6 char**	5 char	4 char	4 char
Age	1	9	**-1**	**1000**	**1001**
Freq. flyer	Y	N	N	N	N
Result	Error	Error	Error	Error	Error

Step 3 – Design

Another example

The tester has to understand what is supposed to happen when a boundary value is entered. When, for example, a field is expected to accept an amount between €0 and €10, does this include €10 or is an error displayed at €9.99? On the other hand, is an amount of €0 acceptable? Are the boundary values in euros or in euro cents? The analyst determines the following boundary values and behavior:

-0.01= rejected
0.00 = rejected
0.01 = accepted
9.99 = accepted
10.00 = accepted
10.01 = rejected

Elements on the same side of the boundary value are put in the same equivalence class, which is why only three values for the upper boundary value need to be tested, namely €9.99 and € 10.00, both of which should be accepted, and €10.01, which should be rejected.

Remarks

- *Determine input values:* In the above example of the online airline ticket store, the same input is used in the name field for a number of test cases. For example, in the test cases 3, 6 and 8 the name has four characters. It is not necessary to use the same name, the input can vary and increase the coverage of the test. In line with the principle of syntax testing, new input values can be determined. As long as all of the values are valid, this shouldn't be a problem. When an unexpected error occurs, the influence of the value concerned must be examined. There may be an error in the input validation.

- *Null values:* A regularly asked question is whether a null value is also a boundary value. It is preferable to handle null values with syntax tests or EP tests. In this example of BVA, the null values are not taken into account.

- *Invalid values:* For a test case with invalid values, the test case should be set up in such a way that there is only one invalid attribute for each test case because it will otherwise not be clear which invalid value triggered the error message. However, this rule does not apply if the system provides an overview of each individual error. If this is the case, multiple invalid values can be included in one test case. Do not forget to take combinations of these values into account; not all combinations may be possible.

Step 3 – Design

- *Difference with EP:* BVA distinguishes between age categories; EP doesn't. The error that a reduction is given for ages between 10 < Age < 60 instead of 10 ≤ Age < 60 will be found with BVA, but not with EP. This makes BVA a stronger technique than EP. Using boundary values in EP instead of arbitrary values taken from the middle of each equivalence class increases the likelihood of finding errors. This has to be done, but it is not called BVA because it does not guarantee that all boundary values are included.

- *Reduction in the number of test cases:* In many cases, the number of tests per boundary can be reduced from three to two values. The assumption is that the boundary value itself is tested and both tests have a different outcome. This is explained in the following example.

It is specified that if the input is > X then Y = B, else Y=A.

Normally, three tests are designed for BVA, namely for input X-1, X and X+1. In this example, X and X-1 produce the same result, namely Y=A. In this case, the test runtime can be reduced by setting up only two test cases: X (this produces Y=A), and X+1 (this produces Y=B).

If the system does not produce the expected outcome, the tests often produce the same value twice (e.g. A-A or B-B). In this case, the third boundary value is tested so that both outcomes A and B have been tested at least once.

12.6.4 C/E: Cause-effect Graphing

Description
A test design technique that uses decision tables to model causes and their effects. C/E provides wide coverage, but is an extensive and time-consuming method [Pol et al, 1999].

Useful for

- Testing small system components with a high risk.
- Assessing the situations to be tested and putting logical test cases on paper. We speak of test assessment when only the approach and no-

tation use the C/E technique. When not striving for completeness and including all possible combinations, C/E is not formally applied. [Mors, 1993].

- Finding ambiguities and gaps in the system design [Pavankumar].

Approach

1. Determine conditions and actions.
2. Define the decision table.
3. Reduce the number of test cases.
4. Define test cases.

Example

Step 1: Determine conditions and actions

Extract the conditions (C1, C2, C3) and the associated actions (A1, A2, A3) from the specifications for a library's member entry system.

When a new member is entered, the system checks whether the applicant is eligible for a youth membership (applicant < 18 years). If this is the case (C1), the young member is added to the family account (A3). If a family account does not exist (C2), the applicant is entered as a new member (A2), but not before the parent's authorization has been checked (A1) and accepted (C3). If the applicant is an adult (applicant ≥ 18 years), he is entered as a new member (A2) or added to the family account (A3).

This produces the two following overviews:

Conditions		
C1: Youth membership		
C2: Existing family account		
C3: Parent authorization is accepted		

Actions		
A1: Verify parent authorization		
A2: Enter new member		
A3: Add member to family account		

Step 2: Define the decision table

The decision table displays all of the conditions and their associated actions. The upper part of the table displays the conditions. The columns are used to display all of the conditions' possible combinations. In this example, the conditions are 1 (true) or 0 (false). The total number of cases is: number of cases = 2^n, whereby n is the number of conditions. The lower part of the table indicates for each case with an X which actions are expected.

Conditions/Actions	1	2	3	4	5	6	7	8
C1: Youth membership	1	1	1	1	0	0	0	0
C2: Existing family account	1	1	0	0	1	1	0	0
C3: Authorization = OK	(1)	(0)	1	0	(1)	(0)	(1)	(0)
A1: Verify authorization			X	X				
A2: Enter new member			X				X	X
A3: Add to account	X	X			X	X		

Note that not all of the combinations are possible or meaningful. In this example, the test cases 1 and 2, 5 and 6, and 7 and 8 cannot be distinguished from each other; they result in the same actions. This is because the condition "Authorization = OK" is not relevant. This is why these conditions are declared with (0) and (1). With C/E, all of the test cases must be run. In practice, this is not always meaningful and the number of test cases is reduced.

Step 3: Reduce the number of test cases

When reducing the number of test cases, the cases with an impossible combination of conditions are discarded, as well as the tests in which the value of the condition has no influence on the result. In the example, we either execute test case 1 *or* 2, 5 *or* 6, and 7 *or* 8.

Reducing the number of test cases narrows the coverage, which is why this needs to be done carefully. There are formal simplification rules that indicate how the number of test cases can be further reduced. These simplification rules are not explained in this chapter, see [Mors, 1993].

Step 4: Define test cases

For each column, define the test cases that contain the combination of conditions and actions in the column.

Remarks

- *The difference with EP and BVA:* C/E tests every combination of the conditions and therefore results in a higher number of test cases. This method is stronger than EP and BVA because these techniques only test one attribute at a time. Possible interactions between attributes are not taken into account.

12.6.5   *State Transition*

Description

A test design technique used to design test cases to test state transitions [Veenendaal, 2002].

Useful for

- Testing process states/state machines.
- Testing screen transitions.

Approach

1. Create the State Transition Diagram (STD).
2. Determine the test depth.
3. Define the state table.
4. Define test cases.

Example

Step 1: Create the State Transition Diagram (STD)

If there are three states and four actions, the STD could look like this:

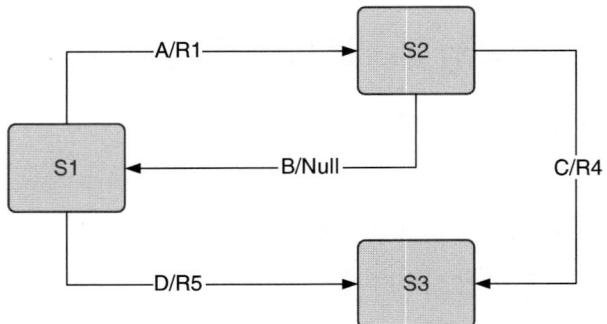

This diagram is read as following:

- In State 1 (S1), execute action A. This will result in a result (R1) and the system will transition to State 2 (S2).
- In S2, execute action B. The system will go back to state S1 without any result.
- An action cannot be executed in State 3 (S3).

A concrete example of this is a library system. A borrowed item is in the state "borrowed" (S1). If the book is scanned when it is returned (A), a receipt is printed (R1) and the book transitions to state "in stock" (S2).

Step 2: Determine the test depth

The test depth depends on the number of steps that are tested each time. A single transition from S1 to S2, for example, is called a first order transition. A greater test depth is achieved when higher order transitions are tested. When applying switch coverage 1, all transitions are second order transitions. From each arbitrary starting state, two steps are taken to reach the final state, for example, from S1 to S3 via S2 (S1-S2-S3) or S1-S2-S1. The switch coverage indicates how many state transitions are tested in each step:

- Switch coverage 0 = first order transitions
- Switch coverage 1 = second order transitions
- Switch coverage 2 = third order transitions

In general, only first order transitions are tested. This example uses STD switch coverage 0.

Step 3: Define the state table

The state table displays the state transitions for zero switch coverage. Since there are three situations and four actions, the state table contains 3x4=12 tests. The below table displays the valid and invalid transitions. Invalid transitions are blank.

	A	B	C	D
S1	S2 / R1	–	–	S3 / R5
S2	–	S1 / null	S3 / R4	–
S3	–	–	–	–

The table is read as following:
In State 1 (S1), execute action A; this results in "R1," the system transitions to State 2 (S2).
In State 1 (S1), execute B; this action should not produce a result; the system remains in State 1.

Step 4: Define test cases

The state table can be used to create transition chains. The following transitions are possible:

S1 → S2
S1 → S3
S2 → S1
S2 → S3

Step 3 – Design

Using each of the above transitions, the following valid and invalid tests can be derived:

	Valid 1	Valid 2	Invalid 3	Invalid 4	Invalid 5	Invalid 5 (cont.)
Precondition	S1	S2	S1	S2	S3	S3
Action	A	C	B	A	A	D
Result	R1	R4	-	-	-	-
Condition	S2		S1	S2	S3	
Action	B		C	D	B	
Result	null		-	-	-	
Condition	S1				S3	
Action	D				C	
Result	R5				-	
Post-condition	S3	S3	S1	S2	S3	S3

Remarks

- The state transition test not only tests state transitions, it also tests the results of the transitions. This means that test 1 will test the results of actions A, B and D.
- Invalid cases can be combined with valid cases. Since no transition is expected for the invalid cases, the system will remain in the same state.
- *Nice to know:* Varying the test depth, as is done in state transition testing, is based on the same principle as the test depth in the Process Cycle Test (PCT).
- *State table and higher order transitions:* In the above example the state table is presented as a 'state by action' matrix. The aim of state testing is to test all transitions. For switch coverage=0 this format works best, the table's columns will list all actions that will lead to a transition. For higher order transitions this format of the state table won't work. In that case use a 'state by state' matrix.

12.6.6   CRUD Testing

Description
Test method that follows a data element in the system or process during its life cycle. The typical life cycle of a data element is defined by Create, Read, Update and Delete. The test is also called data cycle test [Pol et al, 1999].

Useful for
Testing the life cycle of a data element and verifying actions on the data element.

Approach
1. Determine the CRUD matrix.
2. Derive test cases.

Example
Step 1: Determine the CRUD matrix

The CRUD matrix is the starting point for the CRUD test. This matrix can be a component of the system design or it can be derived from the functional specifications. The CRUD matrix specifies in which screens certain actions on a data element are possible. The below matrix is read as follows: the data element "Name" is created (Create) in screen S1, its value is read (Read) in screen S2 and it is deleted in screen S3 (Delete).

	Create	Read	Update	Delete
Name	S1	S2		S3
Age	S4	S2	S1	

Test the life cycle of each data element. The life cycle can be tested for each data element or screen in which actions can be performed on more than one data element. Because the latter is the more efficient option, it is explained below.

Step 2: Derive test cases

Derive test cases by performing all of the CRUD actions in the CRUD table:

	Case 1	Case 2	Case 3	Case 4
Precondition	S1	S2	S3	S4
Action 1	Create Name	Read Name	Delete Name	Create Age
Action 2	Update Age	Read Age		

This table is read as following: Test case 1: In screen S1, enter the element Name (create) and then change the element Age (update).

When tests are designed per screen, the sequence of the test cases is, of course, important. In this example, test case 4 needs to be performed before test case 1. The data element Age needs to be created before it can be changed.

Remarks

- This example includes all of the tests with a positive outcome. The tests can also be extended with negative tests. For example, a negative test will also test that the name attribute cannot be deleted in screen S1.
- When creating the CRUD matrix, the system design needs to be looked at from a specific angle. Creating a CRUD matrix can reveal errors in the design. In the above example, the data element Name cannot be changed after it has been created. This omission can be discovered during the design phase without actually having to run the test. This may, of course, be a deliberate design choice, but it is more likely that it is an error that the analyst can explain.
- If PCT is used to test the organizational procedures, this technique can be easily combined with CRUD tests. The PCT tests run through all of the screens, the CRUD tests can be added for each screen.

12.6.7   PCT: Process Cycle Test / AT: Algorithm Test

Description
Process cycle tests are commonly derived from a process diagram, which can describe the organizational processes (for example, the AO processes), but also a system-technical process [Pol et al, 1999].

This definition combines the algorithm test (AT) and the process cycle test (PCT). Although both techniques have a different scope, the method used to derive the test cases is the same. The algorithm test aims at testing the functional paths in the system while the PCT aims at testing the organizational processes. Users can be involved to test how they will work with the new processes and the new system.

Useful for
PCT: Testing the suitability of the system taking the organizational processes and procedures into account.

Algorithm test: Testing scenarios that run through one or more systems in order to check that the different functions tie in with each other correctly. This test can be run as a quick scan or a chain test if systems will be integrated.

Approach

1. Determine decision points in the process diagram.
2. Determine the test depth.

3. Determine the paths through the process diagram.
4. Define the test scenarios.

Example

Step 1: Determine decision points in the process diagram

Determine the decision points in the process. Define a process diagram
if one is not already available. As previously mentioned, the method
used to derive the test cases is the same for the AT and the PCT. This is
illustrated in the following example. Two almost identical process diagrams are displayed. The left diagram describes a functional path (for
the AT). The right diagram describes a scenario that was created by
linking various use cases (for the PCT). Each block in the diagram is
seen as a statement. Statements can consist of actions (rectangles) or
decision points (diamonds).

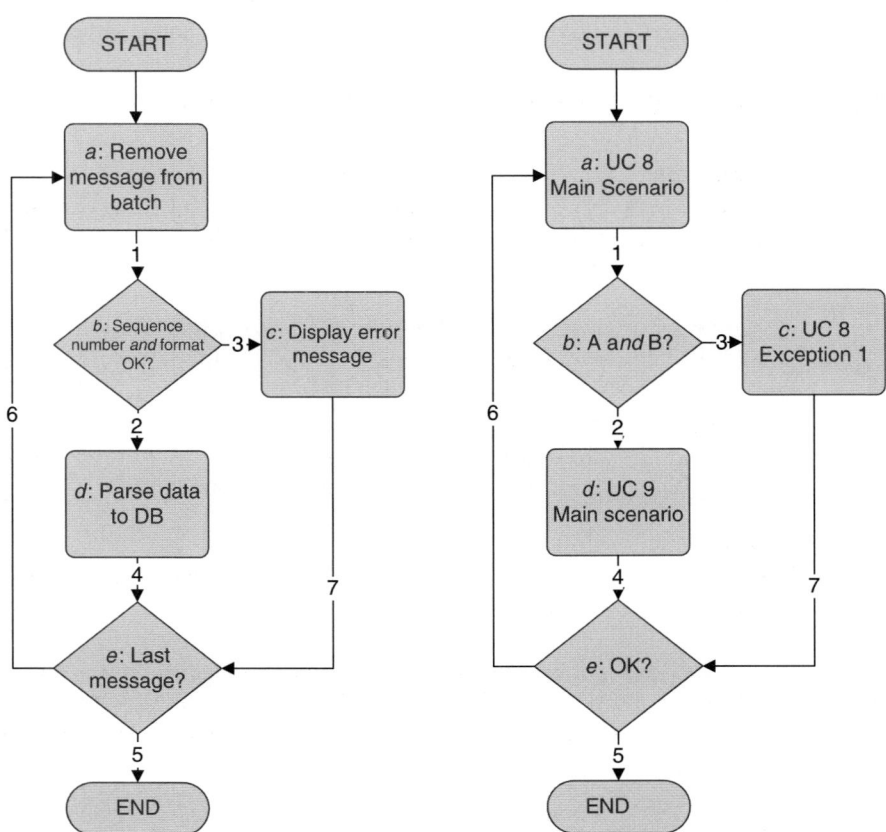

Fig. 12.4 Decision points in a process diagram

Step 2: Determine the test depth

Before test cases can be derived, the test depth needs to be chosen. The test depth for PCT determines the extent to which dependencies between two subsequent decision points will be tested [Computerwoorden]. In practice, test depth 1 is usually used for testing. Test depth 1 means that every path is covered. This test depth has the same coverage as branch coverage. There are two options with regard to branch coverage/test depth 1: increase or decrease the test depth according to the test risk analysis.

If a lower test depth is sufficient, not all results of the decisions will be tested. This test depth is called "statement coverage." With a higher test depth, there is a choice between two types: test depth 2 or "condition coverage." Test depth 2 and condition coverage are not comparable. They are both heavier than branch coverage but focus on different things. Test depth 2 focuses on the flow and tests the path combinations before and after a decision point. Condition coverage focuses on the decision itself and tests the arguments that influence the decision.

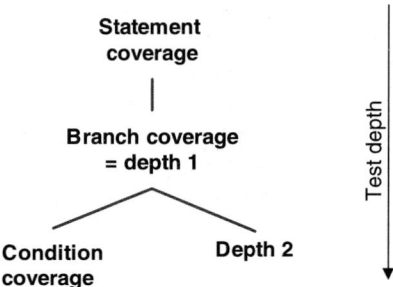

Fig. 12.5 Test depths

To illustrate the above, the test paths for the various test depths are explained below.

Step 3: Determine the paths through the process diagram

Based on the above process diagram, the test paths for all of the test depths are explained in Table 12.2.

Table 12.2 Test paths belonging to different test depths

Statement coverage	All of the statements are run at least once. Two tests can hit all of the statements *a* to *e*. The test paths are: 1-3-7-5 1-2-4-5 Note that path 6 is not covered because it does not contain any statements.
Branch coverage = Test depth 1	Branch coverage: All branches/paths are run once. To achieve branch coverage, paths 1, 2, 3, 4, 5, 6 and 7 need to be run. Test depth 1: In order to test both possible outcomes of the first decision point, paths 2 *and* 3 must be run. For the second decision point, this means that paths 5 *and* 6 must be run. The test cases that cover all of the decisions will also cover all of the branches and vice versa. In this example, this produces one more test case than statement coverage. Because path 6 does not contain a statement, it is not tested with statement coverage but with branch coverage. Branch coverage test depth 1 produces the following test paths: 1-2-4-5 1-2-4-6-1-3-7-5
Test depth 2	All of the combinations between two subsequent paths are tested. These are combinations of arrows *before* and *after* a statement. For the second decision point this produces: 4 – 5 (from 4, to path 5) 4 – 6 (from 4, to path 6) 7 – 5 (from 7, to path 5) 7 – 6 (from 7, to path 6) This produces more cases than branch coverage because there is now a distinction between 4 – 5 and 7 – 5. The outcome of the decision is the same but the starting point is different. Elaborating this idea produces the following two test paths: 1-2-4-6-1-3-7-5 1-3-7-6-1-2-4-5
Condition coverage	All of the conditions determine the outcome of the decision once. The first decision point contains two conditions A and B. If A *and* B are both true, the decision is positive and path 2 follows. With respect to the branch coverage, condition coverage will result in additional test cases in order to cover: 1-2-4 (A, B) 1-3-7 (not A, B) 1-3-7 (A, not B)

In the first test case, A and B are positive because A and B are both true. In the last two cases the outcome of the decision is identical but the condition determining the outcome is different. A is false in the second case and B is false in the third case. According to the definition of condition coverage, the test case in which both A and B are false does not have to be run.

This leads for example to:
1-2-4(A, B)-5
1-2-4-6-1-3-7(not A, B)-5
1-3-7(A, not B)-5

Step 4: Define test scenarios

Two test paths are derived for statement coverage: 1-2-4-5 and 1-3-7-5. In order to cover these two paths, two test cases are defined:

	Case 1	Case 2
Scenario description	Successful processing of batch with *one* message.	Batch processing triggers an error message because the first message has a wrong serial number or format.
Path	1-2-4-5	1-3-7-5

Note that the second test case does not explain the cause of the error message. It is left up to the test engineer who created the physical test design to provide an explanation.

Remarks

- *Nice to know:* The switch coverage used in state transition tests is based on the same principle as the test depth in the process cycle test.

- *Test depth:* As this example shows, a number of essential tests are missing following the use of statement coverage. It is therefore recommended to use branch coverage (test depth 1) for all of the relevant system components. All of the actions are then carried out *once*. Specific system components can then be tested more thoroughly.

12.6.8   Load Tests

Description

Load tests load the system with a representative constant load. The processing times of the time-critical transactions or processes are measured to establish whether they meet the performance requirements.

Useful for

- Determining the (optimal) performance of the system.
- Determining whether further performance tests, like stress tests, are useful.

Approach

1. Determine what the time-critical transactions and/or operating processes are.
2. Decompose the process steps for which the transaction time will be measured.
3. Determine the start and end points of the measurement.
4. Determine the representative load.
5. Design test cases and create the scripts used to run the tests.
6. Run the tests.
7. Analyze the measurement results and aim for a graphical representation.
8. Check whether the results meet the requirements.

Example

Step 1: Determine what the time-critical transactions and/or operating processes are

Determine the time-critical transactions and/or operating processes. The organization may have clear performance requirements, for example:

- A changed customer detail record should be saved within five seconds.
- A management report containing the total revenue should be generated within two minutes.

If the organization doesn't have any performance requirements, determine the time-critical processes. This is best done together with the stakeholders

Step 2: Decompose the process steps for which the transaction time will be measured

A process or transaction often consists of multiple process steps that follow each other after a short wait time. The wait time can be a result of processes that are run sequentially and hence first queued (queue time), or of messages that are sent to other systems (transport time). Decomposing the process into steps clearly illustrates where the real processing time is and where the real wait time is. Figure 12.6 is an example of a decomposition.

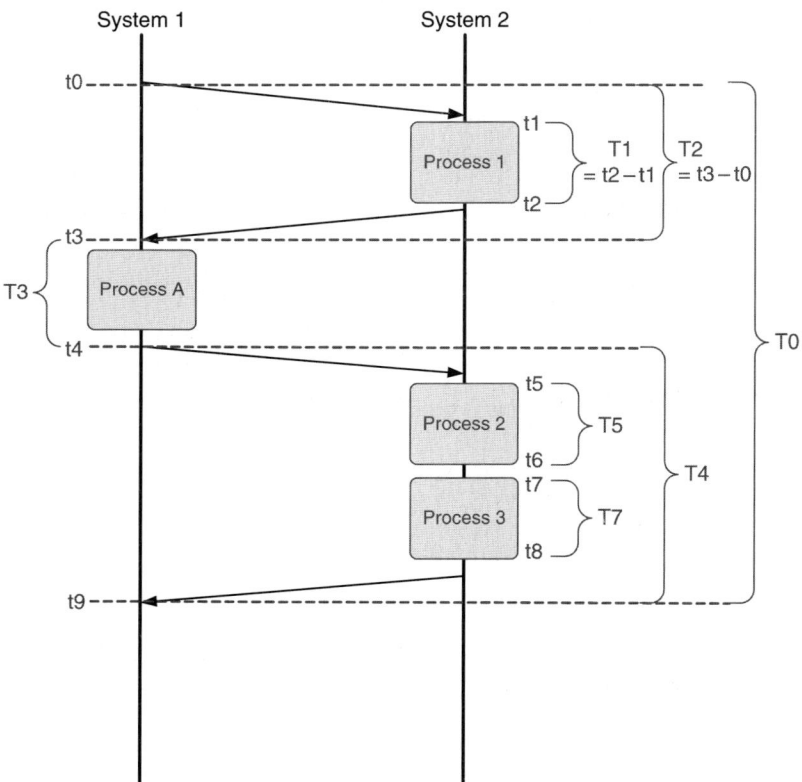

Fig. 12.6 Example of a decomposition

This figure displays the process of two systems that exchange messages. System 1 triggers a message to be sent to System 2, which processes the message and runs Process 1. When Process 1 is finished, a message is sent back to System 1, which processes the message and runs Process A, etc.

When determining performance, it is important to know exactly which time intervals need to be measured. Does the total time of the process on System 2 include or exclude the time required for Process A? Should the processing time for System 1 be left out of the measurement? How is message transport time handled (e. g. t1-t0).

When measuring the processing time of Process 1, 2 and 3, is it a problem that the wait times between the processes are not taken into account (e. g. t7-t6)?

If things are not clear, discuss the design and the expectations with the analyst and the developer. Use this information to ask the stakeholders which time intervals are important to them.

Step 3: Determine the start and end points of the measurement

Step 2 can be used to determine which time intervals are measured and what the start and end points of the measurement are. When determining the measurement points, the technical feasibility is also important. Can the start and end points be determined on the screen, in the log or on the data transport line? If it is very difficult to determine the time-stamp for this point in time, additional measures may have to be taken so it can be measured, for example, by adding additional measuring points to the system logs (sometimes called hooks). If this is too time consuming, select a different measuring point.

For some measurements, a stopwatch can be used to manually measure the processing time. But if the operating processes are complex or if the measurement results need to be accurate, advanced tools should be used (see also Sect. 4.7 Performance Testing).

Step 3 provides insight into the intervals that will be measured and when the time measurement starts and stops. Moreover, step 3 examines whether the measuring points are feasible.

Step 4: Determine the representative load

Load tests load the system with a representative constant load. This load usually consists *either* of a minimum load to measure the maximum performance, *or* of a representative load to measure performance during normal use. The latter requires input from the business or the maintenance organization – they can usually provide information about the expected or the current load of the live system. Assumptions will have to be made if this information is not available.

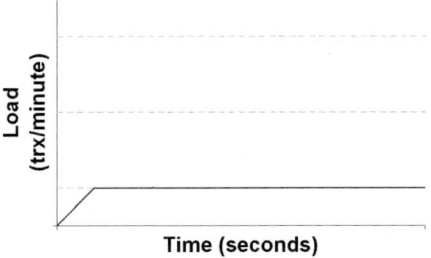

Fig. 12.7 Load testing

The load is usually expressed in the number of users on the system, the number of messages, or the number of transactions processed per unit of time.

Step 5: Design test cases and create the scripts used to run the tests

Now that the load and the performance to be measured have been determined, the tests can be defined. If necessary, scripts are written to run the tests automatically or to generate load. Any tools that are needed for the tests should also be available.

Step 6: Run the tests

Run the planned tests. Every measurement should be done at least three times. If the variation in the measuring values is high, it is recommended to do the measurement more often. If the variation stays high, work with the builders or the analysts to find out what is causing it. Keep a logbook of the measurement and its starting points. It may be necessary to collect the measurement data from different system logs, which is why it's important that you are able to determine which data belongs to which measurement.

Step 7: Analyze the measurement results

When the tests are finished, the measurement results have to be analyzed. The time intervals determined in step 3 are calculated.

The customer will want to know how the system performed. Explaining that a fast transaction time was measured a few times is not enough because it says little about the transaction times that were not measured. Here, we need to apply the probability theory.

The following example illustrates how a basic error analysis can be done. Although this example may be somewhat technical for some readers, it illustrates how probability theory can influence the final conclusion. The example shows that the measured message transport time falls within the defined limit of one minute. However, the error analysis shows that it is realistic to expect that some transactions will exceed the limit.

Example 12.6: Error analysis

The message transport time is measured in order to determine whether it meets the performance requirements. A message should be transferred within one minute. The message transport time cannot be measured directly, so the difference between time intervals T2 and T1 is measured (see also Fig. 12.6). The following measurements are carried out:

Measurement T1	1	2	3
t1	13:51:50	13:59:45	15:16:36
t2	13:55:34	14:03:33	15:20:31
T1 = t2-t1	0:03:44	0:03:48	0:03:55
Average	0:03:49		
Standard deviation	0:00:06		

Measurement T2	4	5	6	7
t0	13:51:12	13:59:03	14:15:59	14:21:02
t3	13:56:03	14:04:00	14:20:30	14:25:55
T2= t3-t0	0:04:51	0:04:57	0:04:31	0:04:53
Average	0:04:48			
Standard deviation	0:00:12			

Time intervals are calculated by subtracting the timestamps of the measurements from each other. The time interval is then determined by the average over the measured intervals:

$$\overline{T_n} = \frac{\sum_1^x T_{n,1}}{x}$$

For the measurement of T1, this produces the following average:

$$\overline{T_n} = \frac{T_{n,1} + T_{n,2} + T_{n,3}}{3} = \frac{0:03:44 + 0:03:48 + 0:03:55}{3}$$
$$= 0:03:49$$

The intervals have a spread that is determined by the standard deviation (σ). Most calculators can calculate this. Excel also has a standard function (formula: =STDEV(B5:E5)) to calculate the standard deviation. Calculating T1 results in $\sigma = 0:00:06$. What does this value mean for the time interval? Elementary statistics assume that 68.2% of all measurement values lie between 0:03:49-0:00:06 and 0:03:49+0:00:06. The consequence of this is that 15.9% of the transaction times are longer than 0:03:49+0:00:06.

The message transport time is determined by calculating the difference between the time intervals T2 and T1. This produces:

$$\overline{T_n} = \overline{T_{n+1}} - \overline{T_n} = 0:04:48 - 0:03:49 = 0:00:59$$

Does this justify the conclusion that the system meets the performance requirement? In order to make a statement about that, the deviations in the measurements have to be studied.

Elementary error analysis shows that for a formula F = A+B-C, the measuring error is equal to the root of the quadratic summed measuring error for every variable, meaning that:

$$\delta F = \sqrt{\delta A^2 + \delta B^2 \delta C^2}$$

Here the error in the time determination (δt) is equal to the standard deviation (σ) mentioned earlier. The message transport time is determined by calculating the difference between the time intervals T2 and T1. The following applies for the error in this measurement:

$$\delta T_n = \sqrt{\sigma T_2^2 + \sigma T_1^2} = \sqrt{0:00:12^2 + 0:00:06^2} = 0:00:13$$

Based on the standard deviation, it can be established that 68.2% of the transport times are expected to be between 0:00:59-0:00:13 and 0:00:59+0:00:13. It is expected that 15.9% of all message transport times will be higher than 0:00:59+0:00:13, and will therefore exceed the maximum time. The probability that there are messages that do not meet the performance requirements is therefore realistic.

A much more extensive error analysis is, of course, possible. The above approach provides a good picture of measuring values and errors in the final result. This approach provides a sufficient degree of confidence unless very strict requirements are imposed.

Step 8: Check whether the results meet the requirements

Testing means comparing the system reaction to the anticipated goal, which is why it is checked whether the measured processing times meet the performance requirements.

If there are no requirements for the system's performance, the acceptability of the performance will be determined in this phase together with the stakeholders. If the performance is too low, the cause will have to be determined and the problem solved.

If the performance is sufficient at a low load and it is not yet known how the system will react at a higher load, it is useful to run stress tests to examine how the processing time changes when the system load is increased.

Remarks
Difference with stress tests: Load tests measure performance at a constant, representative load. The aim of the load test is to measure the (daily) processing time. Stress tests mainly focus on the effect an increasing load has on the system's processing time and on situations in which the system grinds to a halt. The processing time is measured to establish its load dependency.

12.6.9   Stress Testing

Description
Stress testing consists of loading the system with an increasing load. Stress tests determine the load under which a system fails and the way in which it fails. A stress situation can crash the system, for example, but can also cause the system to react very slowly. There are several kinds of stress tests, such as step stress tests or peak stress tests.

Useful for

- Determining the maximum load the system can bear.
- Determining the influence of load on the processing time.

Approach
The approach is the same as the approach for load testing, with the following additions.

Step 4: Establish the maximum load

Instead of the representative load, the maximum load is determined here. The maximum load is often expressed as a constant load and a peak load. For example:
"The system should be able to record 200 transactions/minute, and each transaction should be finished within (at the most) 1.0 seconds. The peak load is 10 transactions per second, whereby the peak load lasts for a maximum of 2 seconds. In this situation the transaction time does not exceed 1.5 seconds."

Step 5: Design test cases and create the scripts to run the tests

Design test cases that can establish the breaking point. Create the scripts to run the tests. In some cases, load tests can be run manually, but stress tests are virtually impossible to run without tools. To increase the load on the system, users are sometimes asked to perform an action at the same time. This can be practical, but with this type of tests there is little control over the load, making them difficult to reproduce.

Step 6: Running the tests

Run the tests and check whether they meet the requirements for response time, breaking point, system recovery and security.

The initial load is increased during the test. This is usually done gradually, so the system can find its "balance" before the performance is measured. An exception is, of course, the peak load.

Example
Figure 12.8 displays the load characteristics for stress tests with a gradually increased load and a peak load.

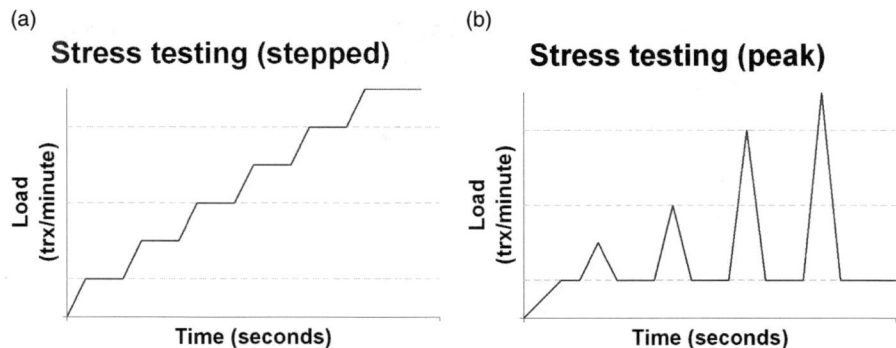

Fig. 12.8 Load characteristics for stress tests

Gradually increased load
The load is increased and kept constant for a while. The system behavior is monitored in order to determine how it reacts to a load that exceeds the set threshold over a longer period of time.

Example 12.7

During a stress test, the system stops starting new sessions once the load has exceeded a certain threshold. The maintenance organization receives a message indicating that the load has exceeded the maximum allowable threshold. New users cannot log in. Existing sessions are processed without notable delay. Because the observed behavior meets the system design, the test is successful.

Peaks in the load
Peaks are brief system loads that exceed a specific threshold. The system either recovers correctly and processes the backlog after the peak is over or it does not recover and unexpectedly shows undesired behavior. If such behavior occurs, the cause will have to be determined and solved.

12.6.10 Reliability Testing

Description
Reliability tests test the reliability (or stability) of the system by making the system operate under a stable, representative load over a longer period of time. This test type is also called an endurance test.

Useful for
Determining the stability of the system.

Approach
The approach is the same as that for load testing, with the following additions:

Step 1: Determine the time-critical transactions and/or operating processes

Load testing mainly focuses on measuring the processing time of time-critical transactions or operating processes. Reliability testing mainly focuses on the stability of the system. A simulation of the usage of the system shows whether the performance decreases during use, for example, as a result of memory leaks. Performance is usually measured by running a few transactions and/or operating processes.

Step 6: Running the tests

Reliability tests take long and are run automatically, often at night or on weekends. If an endurance test runs without problems for a night or a weekend, the duration of the test can be extended.

Example
The load characteristics for reliability tests with a long-term constant load are displayed below:

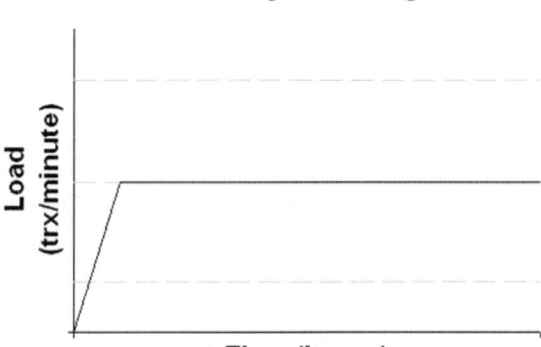

Fig. 12.9 Load characteristics for reliability tests

Remarks
Don't start the endurance test when everyone starts leaving for the day; start it a few hours earlier. If something does go wrong, the problem can probably be solved and the test restarted. Experience shows that reliability tests often terminate within an hour of starting because of little mistakes or errors such as a wrong configuration. These errors are generally easy to fix. Starting the test earlier in the afternoon prevents you discovering the next morning that the test stopped five minutes after the last tester went home.

12.6.11 Concurrency Tests

Description
These tests look at the correlation of user actions on the system. These tests aim to determine whether concurrent transactions have a negative impact on each other.

Useful for
Determining the behavior of the system when several transactions are run simultaneously.

Approach

1. Establish which processes can influence each other.
2. Determine scenarios.
3. Design test cases and create the scripts to run the tests.

Example
Step 1: Determine which processes can influence each other

For these tests, it is important to establish which processes can influence each other. These tests require the assistance of people who have a (technical) overview of the application to be tested, and who can point out where the difficulties in the architecture and technique are so they can be tested.

Step 2: Determine scenarios

The processes established in step 1 will be developed into test scenarios. There are two options:

1. Steps from the processes are run sequentially. For this option, a number of steps are performed manually or automatically, whereby mul-

tiple users are simulated. This option works well if the direct connection between the actions is clear. For example:

User 1	Administrator
1 Logs in	
2 Goes to screen x and enters data	
	3 Logs in
	4 Goes to user administration
	5 Removes the permissions of user 1 in screen x
6 Selects "Save record"	

In this test case, a correct system reaction would be that User 1 can finish what he was doing. Another implementation could be that the administrator cannot change user permissions while the user is logged in. The error that is looked for is the error the system displays when the user saves the record.

2. The processes are repeatedly run at the same time. This option is especially suitable if it cannot be estimated when and in which way the processes influence each other. By executing the processes repeatedly and simultaneously, undesired interaction may be discovered. Tools are needed. Besides having to generate the load, the tools should also be able to check the system's reaction. The system's reaction will show when the processes start interacting. The tools should then be able to establish which process triggered the system's reaction.

Step 3: Design test cases and create the scripts to run the tests

Define the scenarios and the way in which they have to be combined. If necessary, scripts are written to execute the tests automatically. The required tools are developed.

Remark
Performance measurements: "Concurrent" use can also bring performance problems to light. Problems can occur if certain processes have a negative impact on the performance of other processes. To find these errors, the processing time will have to be measured as well.

12.6.12 HT: Heuristic Testing

Description
This is a test design technique that uses existing knowledge and experience. This knowledge and experience is stored in checklists that can be

reused for future tests. The checklists tell the tester what he should check, and force completeness. Heuristic testing can be used for dynamic tests and static tests.

Useful for
Tests that can be reduced to checking specific things using predefined checklists.

Approach

1. Use the available checklist.
2. Evaluate the contents of the checklist (is it suitable for my purpose?).
3. Check the product using the checklist.
4. Evaluate the checklist and adapt it. Store the acquired knowledge so it can be used next time.

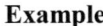

Example

Static testing: In TestGoal, checklists are used, among other things, for the sanity check, the smoke test and to develop the requirements for the test environment. See the relevant chapters for a description of the use.

Dynamic testing: If certain components need to be checked frequently, it can be useful to not describe them in detail every time in the physical test design. This would result in a very voluminous test design with a lot of duplication. It can be practical to put the standard checks on a generic checklist. An example is the testing of the application's GUI. The checklist can show which elements need to be tested on each screen: The navigation, the log-off link, the pull-down menus, etc. These lists can be made for each test, but generic checklists are also available, such as the SUMI list to check user friendliness.

Remark
TestGoal contains a lot of checklists, but other methodologies often contain many standard lists, too.

12.6.13 ET: Exploratory Testing

Description
An approach for unspecified tests that is based on the skill and experience of the testers. ET is a formal, risk-based technique that uses procedures, test charters and heuristics [Veenendaal, 2002].

Step 3 – Design

Useful for

- Testing off the beaten track.
- Testing if no time is left to write test scripts and the risk of the function is not very high.
- Testing if few system specifications are available, but knowledge of the system and the anticipated goal is.
- As a supplement to the "scripted" test design techniques.

Introduction

The philosophy of ET is that the experience gained during a test is immediately used to determine what the focus of the remaining tests will be. This is in strong contrast to the other test design techniques, where test cases are developed in advance. Testing with predefined test cases is called "scripted testing." Writing the tests down basically defines which tests will be run. Favorers of ET say that it is inefficient to predefine detailed tests. The arguments are twofold.

1. The added value of the tests can be derived from the number of solved errors. The time that is normally spent on the test design is better for hands-on testing and error detection.
2. During the test, the tester discovers the system's real weak spots. Running predefined tests has little added value if it has already been determined that the problems are of a different category. In this case it is better to create new tests that cover the category. But at this point there's usually no time left to add a complete new test set to the test design. So an "unscripted" approach is needed.

Exploratory testing is an unscripted test design technique. Test cases are not developed in detail. However, a tight process and clear work packages guarantee the right choices are made and that what has and has not been tested is traceable. The process is supervised by the moderator, who may be the test manager or the test coordinator [Veenendaal, 2002].

Process

1. Putting together the test team

In ET, tests are run by teams of two testers. When putting the team together, the moderator must ensure that each team has knowledge of the test and business domain. This can be done by choosing two experienced testers with knowledge of the domain, or by creating teams of testers and senior users.

2. Organizing the kick-off

In ET, the observance of the agreed process and the mindset of the people involved are very important. This is why the moderator organizes a kick-off before testing starts to ensure that everyone involved understands the philosophy and the process of exploratory testing. When everyone understands what is expected of them, the work packages (test charters) can be defined.

The process shown in the below figure is used [Veenendaal].

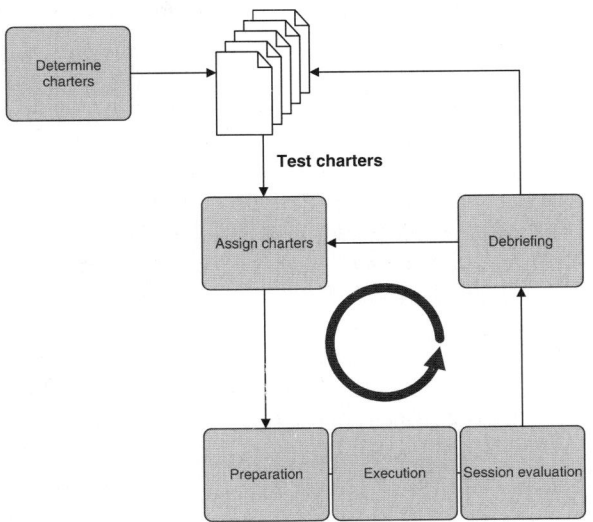

Fig. 12.10 Exploratory testing

3. Determine the test charters
All the test activities are done according to test charters . Test charters are work packages of two hours to one day. The test charter tells the tester what the scope of his test is and what he needs to pay attention to. The test charters are defined and prioritized together with the stakeholders. This process resembles the TRA in many aspects (Chap. 7 Test Risk Analysis). The result of the session is a list identifying the test charters that need to be developed and their relative importance. The test coordinator ensures that this list is included in the test charters before the first test session is held.

Among other things, a test charter contains the following information (see the template in Appendix C):

Charter ID
Unique reference.

Priority
The relative importance of the test cluster.

Time Available
The time that can be spent on the charter.

Anticipated Goal
The goal the charter is aiming for, for example, "Demonstrating that it is possible to send a mailing to subscribed customers."

Why Test
Why it is important to test this charter. In fact, this is a link to the business goal, which describes the contribution the tested function makes to the goal. Including this link helps the testers validate the implemented solution.

Expected Problems
If known, the identified risks or possible problems that are associated with the tested function. The testers use this knowledge to decide which tests they will run.

Do Not Test in This Charter
Things that should not be included in a charter, for example, because a separate charter was defined for these areas of attention. For example: "Do test the sending of the data, but not the correctness of the sent data."

Conclusion
The conclusion of the charter, which is written after the charter has been executed. The conclusion establishes whether the charter is finished (if this is not the case, it may have to be continued in the next session); should be extended or whether a new charter should be defined; has achieved the anticipated goal.

If the charter does not have a positive conclusion, the errors are specified. They can be the errors that were entered in the bug tracking tool. However, if several errors were found, it is clearer to explain what the essence of the errors is by describing them and the risk involved and including a proposal for a suitable measure.

ET Test charter: 21 PROD_Send   **Prio: C**

Conclusion

Conclusion	Y / N
The charter is finished.	N
No extensions to this charter are required.	Y
The anticipated goal of this charter has been achieved, the risk is sufficiently covered.	N
Can we go live based on this charter?	**N**

If the conclusion is negative:
Indicate which action is necessary to achieve the charter's anticipated goal. Indicate which risks have not been sufficiently covered and provide, if requested, a proposal for new charters

Findings	Risk	Activities of anticipated goal of the proposed charter
Gamma: 720M: A new shipment cannot be saved	The customers will not receive any information	See issue 1452

Fig. 12.11 Part of a finished test charter

4. Assigning charters
The moderator sorts the charters by priority and assigns them to the test teams according to their skills.

Because some charters are more exhaustive than others, the amount of time that should be spent on a charter, for example, 2, 4 or 8 hours is specified on each of the charters. When assigning the charters, the moderator makes sure that each team has enough work until the next debriefing. For example, one team can be assigned two charters of 2 hours, while another team is only assigned one charter of 4 hours.

5. Run the test (preparation, execution, session evaluation)
The test team first studies the charter. They examine the charter and discuss how they should run it. The team establishes what has to be tested, what the general approach is, who will run the test, and who will be the "scribe." A test charter of 4 hours requires about half an hour of preparation.

The tests are then run. Both testers are supposed to challenge each other and stimulate each other to create interesting test cases. One person usually runs the test, while the other, the scribe, makes notes in the test log. The advantage of a test log is that the tests that were run can be traced at a later stage. During the session, all of the errors are entered in the bug tracking system. The time also needs to be monitored during the test. In principle, the testers can decide how they run the test charter. If a lot of errors are found in a specific function, it makes sense for the testers to spend more time on it. It is also important, though, that all of the areas of attention in a charter are dealt with. In order to do this, a

Step 3 – Design

scan is run on the whole charter first, the so-called "mile wide, inch deep" test. The testers use these first test results to determine where problems are expected to occur, and concentrate on those categories. This approach prevents large parts of the test being left out and enables the team to draw up a conclusion at the end of the session.

The test teams need to work in a quiet environment. This means that they should work in a closed room and that all their other tasks should be assigned to colleagues.

The last fifteen minutes of the session are used to prepare the debriefing. Together, the test team defines

- The conclusion of the test charter
- Recommendations for possible future tests
- The assessment of the quality
- The most interesting bug that was found

> Experience shows that it is effective to use a flip chart. Each team has five minutes to present its results to the other teams. This way of working gives the debriefing a good structure and forces the testers to be well prepared.

6. Debriefing

The debriefing is led by the moderator. The aim of the debriefing is for the test teams to exchange experience. The testers discuss the results and possible risks of the finished session and exchange ideas for new tests. If necessary, new charters are defined and prioritized, and the charters for the next session are assigned.

The debriefing should be limited. Aim at spending no more than fifteen minutes on each charter. It's important that the moderator and the testers are well prepared for the debriefing so that everything can be discussed quickly. Experience shows that a lot of time is wasted on the debriefing if the meeting is not structured. This strongly decreases the effectiveness of ET.

The described process is repeated until all charters have been tested or the agreed time is over.

7. Retesting and regression tests

In principle, retesting is not done according to the exploratory testing technique. The errors are logged and provide enough information to retest. The essence of the ET sessions is the exploration of the test object and, in particular, the search for new errors. It's important that no time is lost retesting the errors. Showstoppers are retested, but it is preferable to postpone the remaining retests until after the ET sessions. Because it is not necessary to retest in teams of two, the retest can be run when not all testers are available.

After the errors have been retested, the regression tests starts. Because the testers kept a log, the earlier tests can be repeated. The degree of improvisation depends on the detail of the test log. Note that if a high reusability is desired, there are high demands on the test log.

Areas of Attention
Just as for the other techniques, exploratory testing is very effective in some situations and not in others. For example, it may be desirable to work from a predefined test design if the risks are very high and if you want to be sure which tests are run and how they're run. Scripting may also be desirable if the tests are supposed to be reused. Both of these arguments apply to conformance testing. Not only do all of the aspects of the standards have to be tested before it can be established that the test object meets them, the tests are usually run on a large number of different test objects. Exploratory testing is not really a suitable technique in this situation.

Exploratory testing also becomes less effective when it is difficult to validate the system reaction in real time, which can be the case with complex calculations. The time that is needed to establish that the calculation is correct takes the dynamics out of the test process. In this case it is more efficient to specify the expected result in advance using predefined test data. The system takes too long to provide feedback for batch processes with a long runtime to apply ET effectively.

ET also puts demands on the testers. Good testers are needed, who can come up with good tests on the fly and run them. The testers have to be able to deal with uncertainties, be creative, and be so familiar with all of the test design techniques described in this chapter that they can run them with their eyes closed. They also have to have knowledge of the business domain and have a feeling for finding errors. Exploratory testing has a playful element, which makes exploratory projects challenging. But this is only true if the tester can appreciate it.

Don't forget to account for the overhead that will be generated, for example, by the debriefings, the definition of new test charters, and the updating of reports and lists, such as error reports and the list of test charters. These overviews are necessary to monitor the progress and quality during the debriefings and should be updated daily. Make sure you schedule enough time for these activities.

12.7   Test Design Techniques and Security Testing

A security test aims at protecting IT systems. Security tests can focus on IT components such as DMZ servers, firewalls, router configurations and standard workstations, as well as on a complete network or application infrastructure. A system security test is particularly suitable to analyze systems with a high risk profile for security problems. Security tests, too, require a tight structure. The test design techniques described in this chapter are very useful for security tests. Security tests don't usually take the "normal" way of using the system into account. A good security tester will test a server (like a Web server) using the client (like a browser), but will also access the server through the back door, just as a malicious user will.

Table 12.3 shows how the techniques can be used for structured security testing:

Table 12.3 Techniques for structured security testing

Test design technique	Specific to security testing
Syntax testing	A large part of the security problems result from errors in input validation. This is why syntax tests are often run with especially long inputs or input with specific symbols.
Equivalence partitioning	Values in the equivalence classes that lead to negative or very large numbers (32-bit overflow) can cause security problems, which is why the invalid equivalence classes are included. Also think of the output EP, which enables us to test situations in which two valid inputs produce an invalid output.
Boundary value analysis	In particular boundary values that lead to negative or very large numbers (32-bit overflow) can cause security problems.
Cause-effect graphing	C/E can be used to do a test assessment. This technique is suitable for finding ambiguities and gaps in the system design. Where there are gaps, there may be security leaks.
State transition testing	This technique has no specific application in security testing.

Table 12.3 Continued

Test design technique	Specific to security testing
CRUD tests	CRUD tests are suitable for checking authorizations: for each authorization or user profile, the test checks whether it is impossible to carry out prohibited actions. For example, a requester who cannot start the function that the request approves, or a user with read-only permissions that can suddenly change data. Security tests mainly use the negative tests in the CRUD table.
Process cycle testing	This technique has no specific application in security testing.
Exploratory testing	Very suitable for security testing. This technique is often used to flexibly exploit the discovered security leaks. ET combines the flexibility of the test run with a tight process to produce a repeatable and controlled process that takes the risk analysis into account.
Heuristic testing	The use of checklists to test for the presence of specific security heuristics is very important to achieve a reproducible and high-quality result.
Load testing	This technique has no specific application in security testing.
Stress tests	This technique is used in security testing to check whether the system is still in a secure mode when the load exceeds the "breaking point" when the system can crash or go into overload mode. In either case, the system needs to stay at the same security level.
Reliability testing	This technique has no specific application in security testing.
Concurrency testing	This technique has no specific application in security testing.

Step 3 – Design

13   The Physical Test Design

13.1   Introduction

This chapter describes the creation of physical test cases. The starting
point for the creation of a physical test case is the logical test case. The
tester uses his knowledge of the system to convert logical test cases into
physical test cases, which define the test actions that need to be carried
out, the test data used, and the most efficient sequence (the testing scenarios) to carry them out in. These three elements form an integral
whole because they are all worked on during the creation of the physical test design.

If a logical design is not available for a specific test cluster , the test
base needs to be directly converted into physical test cases. This is possible, but it has the disadvantage that it will be difficult to make a good
statement about the test coverage.

13.2   Relationship Between the TRA and the Logical
Test Design

A good structure is very important for a physical test design. A test design for a medium-sized system can easily contain a few hundred pages
and more than a thousand test actions. A good structure and clear references to the test base enable tests to be traced and increase maintainability. The basis of all test actions, and thus of the physical test design,
is the TRA. During the creation of the TRA, a test tree is set up. This
test tree is the basis for the structure of the physical test design. Giving
all of the tests a position in the tree creates a hierarchical structure in the

Step 3 – Design

D.-J. de Grood, *TestGoal*,
DOI: 10.1007/978-3-540-78828-7_13, © Collis B.V., Leiden, The Netherlands, 2008

test design, which ensures that a test can always be traced to the corresponding risk and component in the test base.

Figure 13.1 displays the relationship between the TRA and the logical and physical test designs. The TRA is done using the test tree. The importance is estimated for each branch in the tree. The test tree specifies the clustering of the tests that are to be designed. Every cluster is defined in logical and physical test cases. Using these structures ensures that the result of a test action that has been performed can always be traced to the area of attention discussed during the TRA.

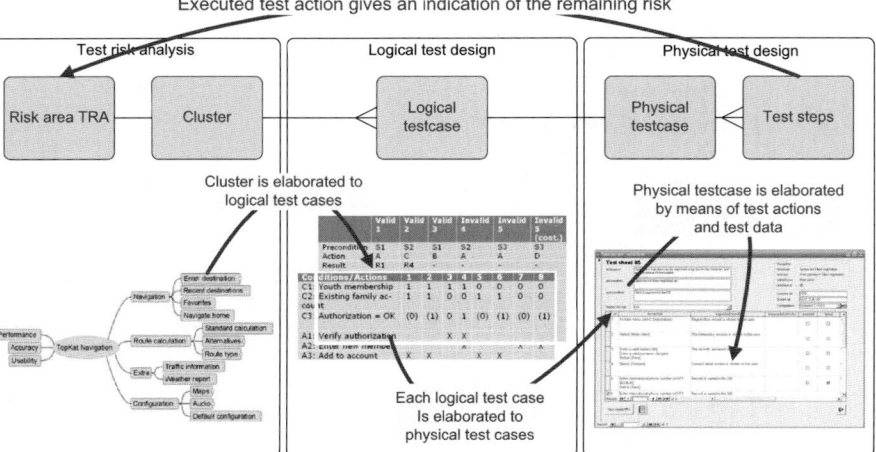

Fig. 13.1 The relationship between the TRA and the logical and physical test designs

13.3   Physical Test Case

The logical test case describes *what* is tested and not *how* it is done. The aim of a physical test case is to fill in the details of the corresponding logical test case so that the *how* is clear. In order to elaborate the physical test case, we need to answer the following questions:

- Which preconditions or other conditions need to be met?
- Which actions need to be carried out?
- Which input data should be used?
- When is a test successful?

A physical test case consists of the following attributes:

Physical Test Case ID
Each physical test case has a unique identifier, which can be a number or a text. A text may be an easy way of quickly identifying a test, but a number, like in the first of the below examples, is easy to generate automatically.

Possible examples of IDs:

- **Test Case 1**
  The first test that was designed.
- **CC_auth_nok_pin_01**
  This test case ID specifies that the test case is about an invalid (nok) credit card (CC) authentication (auth) with an incorrect PIN.
- **CC_auth_nok_blacklist_03**
  This test case ID specifies that the test case is about an invalid (nok) authentication (auth) and that the credit card (CC) is on a blacklist.
- **INT02.C1T01**
  This test case ID has the format <project><cluster>.<condition><test case>. In this case, the ID stands for the second cluster, condition C1, first test case.

Test Purpose
To provide a short description of the test case, which corresponds to the logical test case and clearly indicates when a test case can be called successful. The description should include the following elements:

- The situation that is tested (for example, "enter an invalid date")
- The expected and correct system reaction (for example, "cannot save the record")

This produces descriptions such as:

- Check that a record is not saved if the date is invalid.
- Check that the authentication is not approved if a wrong or invalid PIN is entered.

Precondition
Describes the test's base situation and specifies the conditions that must be met when starting the test case. For example, in the case of a library system: *System in window S1, status of the book = on loan.*

Step 3 – Design

> Describe the standard, valid base situation in a central place in the physical test design. In the test case, all you then have to do is specify where it deviates from the valid situation. This prevents the precondition turning into a long list of obvious points that provide no information about the actual test with conditions such as: the user is logged in, the connection with the database has been established, the database contains a number of book titles, etc.

Only specifying where the test case deviates from the valid situation provides better insight into the essence of the test. For example, in the valid situation, the book status is always available, but in this test an exception is made: *status of the book = on loan.*

> Using mathematical symbols such as $=$, $\leq$, $<$, $\geq$ and $>$ for conditions ensures clarity. This notation is often more precise than a textual description and it stands out in the text. This makes recognizing the condition and looking for the differences between various test cases easier.

Test Actions
A test action consists of one of the following attributes: test action ID, test action, expected system reaction, conclusion and the indication for regression testing. These attributes are discussed in Sect. 13.4.

Post-condition
Describes the expected situation after the test. The post-condition is described in the same way as the precondition.

Reference to the Test Base

Specifies the part of the test base the test is based on. This makes it easy to locate the related specifications.
This field is important to locate the specifications for review and control.

Reference to the Position in the Test Tree
Specifies the cluster or logical test the physical test case belongs to.

Quality Attribute
Specifies the quality attribute that is tested by the test case. This field enables the test results to be displayed in the test report by quality attribute.

Reference to the Configuration Used

The tests are run with a specific configuration of the test object, test environment, test base and test data. Specifying the versions that were used to run the test enables the result of changes to be traced.

> The configuration can consist of different kinds of elements. Experience shows, that it is convenient to describe these elements in the configuration management system. The unique combination of specific versions of test object, test environment, test base and even test data, enables us to assign a unique label, namely the configuration ID, which is referred to in the test cases. The configuration ID can be chosen freely, but it is advisable to use a name that is convenient and clear for the test team.

If one or more elements in the configuration change, the description is updated with the new configuration and the difference between the previous and the new configuration noted. The configuration ID of the new configuration is used as reference in the test cases. The configuration description can contain a reference to the used:
Test object
The version number of the test object is used to determine which errors have been fixed and can be tested or retested. The release notes indicate which errors were fixed and which functions are new to a particular version.

Test environment
Not only the test object is subject to change, the test environment is too. Examples of such changes are database migrations, patches and updates to third-party components. It may not be necessary for the test team to manage all the changes, but they do have to manage the changes that can impact the tests.

Test base
The tests are designed for a particular version of the test base. There is not always enough time to implement changes to the test base in the test design, which is why you should always specify the version of the test base the test design is based on. This enables you to determine which changes in the test base still have to be implemented in the test design and to explain failed tests. If the programmer says the system is working correctly but the behavior deviates from the expected outcome, it is possible that changes were made to the test base that were not implemented in the test design.

Step 3 – Design

Because the test base often consists of a large number of documents, it's a good idea to compile a list of the documents and give the list a version number. When system specifications are released as a new "baseline," a new list with a higher version number can be used as the test base.

Test data

The input and output data that is used in the physical test case depend on the configuration and metadata (see also Sect. 16.13.6). Any changes to the configuration will impact the usability of the test cases, which is why you should specify which version of the test data was used for the designs.

Write the configuration on a whiteboard during the test run so that everyone in the room knows which configuration is used for the test environment. If the testers are not working in the same room, an Intranet page or a wiki web are also good places to publish the configuration.

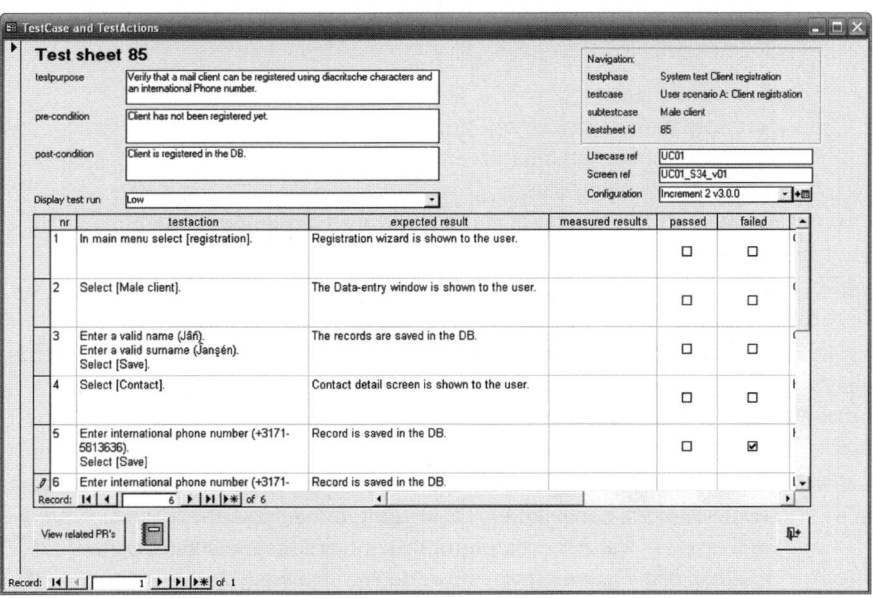

Fig. 13.2 A test case in Microsoft Access

Figure 13.2 displays an example of what a physical test case looks like in Microsoft Access. The above-mentioned attributes are recognizable, as well as the test actions that describe the actual test steps. The test actions are described in Sect. 13.4.

13.4   Test Actions

A physical test case consists of one or more test actions . A test action consists of the following attributes:

Test Action ID
Unique identification of the actual action the tester has to perform, preferably a number that is unique in the physical test case. The combination <physical test case ID>.<test case ID> should be unique so there is no confusion.

An example of unique IDs for the first three test actions of test case 4:

- Test case 4.1
- Test case 4.2
- Test case 4.3

Test Action
Describes an action that the tester has to perform. Use action words such as

- Select
- Check
- Create
- Enter
- Edit
- Delete
- Log in

If the physical test case is used as a basis for automated test output, it is recommended to use the same action words in the test scripts and in the physical test case.

Also specify clearly which input data needs to be used. Experience shows that it is convenient to specify if valid or invalid values are entered because it makes the test easier to understand. For example:

- Enter an invalid date (31-02-2005).
- Select an invalid user (Mr. Johnson).

Chapter 14 Test Data provides a number of examples of how test data can be specified in physical test cases.

Expected System Reaction

The expected system reaction describes the result that is expected if the system behaves according to the design. Describe the result in such a way that it can be checked, for example, by mentioning the action the system performs and for whom the action is performed. For example:

- *The system displays an error message to the user to indicate that the date is invalid.*
- *A reservation request is sent to the library system.*
- *The result of the computation (answer = 42) is displayed to the user.*
- *The changed book status is written to the database.*

Conclusion

Specifies whether the action was successful. A test action is successful if the observed result is the same as the described expected result. The value of the conclusion can be OK, NOK, or not run. If a test was not successful, a reference to the associated error is added.

Sometimes, an additional status is added to the conclusion to indicate that a test was not feasible. The disadvantage of this, is that a non-feasible test is often deemed "not important anymore" in the progress meeting. If a test is not feasible, decide whether it's important, and if it isn't, delete it from the test set or the regression test set. If, however, the test is important, the fact that it has not been run is an open risk. Leave the test in the report as "test yet to be run."

Indication for the Regression Test

Indicates the risk category a test action belongs to. This ranking by importance can be used to determine the regression test set.

A distinction can be made between full regression tests and, for example, quick scans. Some urgent changes need a quick and limited regression test. Indicating the importance of a test action makes it easier to define a regression policy; see also Table 13.1.

Table 13.1 Regression test policy

Type of regression test	Run
Quick scan	Only run tests and retests in the risk category "critical."
Limited regression test	Only run tests and retests in the risk category "critical" and "high."
Full regression test	Run all tests.

Depending on the situation, this indication for regression tests can also be established at the test case level. In practice, it requires less detail. An advantage of an indication at the test action level is that in a physical test action case, the tests that are not strictly necessary can be left out. This can, for example, be a syntax test where different valid values are entered. In an exhaustive test we would want to perform this test completely, but not in a quick scan. Reducing the number of tests changes the coverage but not the way in which the test case is run through.

13.5   The Physical Test Scenario

The physical test design consists of a number of physical test cases. The sequence in which they are added to the test design is determined by the test tree. The advantage of this structure is that test cases are easily traced and the test set is easier to maintain.

A more efficient sequence can generally be devised for the test run. This is done by creating a separate test scenario that specifies the sequence in which the test cases are best run. This is generally the logical sequence in which two test cases immediately follow each other. For example, the first test case creates a record and the one that follows changes the record. If the post-condition of the first test does link up with the precondition of the one that follows, link actions are defined to make sure the tests can be run immediately after each other.
A scenario consists of the following attributes:

Scenario ID
Unique identification of the scenario.

Description
Describes the scenario in a concise and clear way. Make sure the stakeholders can create a picture of the scenario. The test goal in the physical test case is meant for insiders; the scenario will have to tie in with the experience of the users, the marketing department, the program directors etc.

Tests to Be Run
A list of the tests that are to be run in the scenario supplemented with possible link actions. For example:

- Test case number 1
- Test case number 5

- <link action>
- Test case number 3
- Test case number 7

Conclusion

The conclusion of the scenario. The value of the conclusion can be OK, NOK or not run. Some scenarios can be run through and completed, and still contain a number of unsuccessful tests. If this doesn't produce any showstoppers, for example, because there is a "workaround," it can be worth giving the scenario an OK conclusion. Accepting the scenario indicates that the business process is feasible.

A reference matrix like the one displayed in Table 13.2 provides for good traceability. This cross-check matrix shows which physical test case is tested in which physical scenario. The matrix is primarily used to control completeness, but it also increases the reusability of the test design. It's best to keep the matrix up to date.

Table 13.2 A reference matrix

	Scenario 01	Scenario 02	...	Scenario nn
Test case 1	X			
Test case 2				X
Test case 3	X			
Test case 4		X		
Test case 5	X			X
Test case 6		X		
Test case 7	X			

13.6 Test Data

While setting up the physical test design, the test data used during the test is also defined. The various types of test data and the way they are used in the physical test design is described in Chap. 14 Test data.

14   Test Data

14.1   Test Data Elements

People often think of test data as the data that is input and checked when a physical test case is run. But test data covers much more than that. The following elements are distinguished:

Input and Output Data
During the test run, data is entered and the output of the test action is compared to the expected result (expected output data). Input data is data that is defined for boundary value research, syntax tests, etc.

Output data can, for example, be the output of a calculation, or data elements that are generated by the system for messages and reports.

Operational Data
Each test has a precondition. During the test, the data set can be searched for data or processes that are in a condition that meets the precondition. The data that is present in the system and that can be used as the starting point for a specific test is called operational data. This name was chosen because this kind of data is also present in the live system where "operations" are carried out.

The base situation described in the test design can be created during the test run, but it is more efficient to create it prior to running the test and to make it available in the test object's database so it can be used when the test is run. The base situation often consists of a process that is in a specific state, for example:

- A specific credit card that is also on a blacklist
- A library system in which a specific book has the status "on loan"

D.-J. de Grood, *TestGoal*,
DOI: 10.1007/978-3-540-78828-7_14, © Collis B.V., Leiden, The Netherlands, 2008

Configuration Data

One of the requirements of testing is that the system is correctly configured. The configuration data is the data that is used to make the system suitable for use, which is why it is important that the correct configuration data is used. Configuration data can vary depending on the test level. The following questions should be answered to ensure the system is correctly configured:

- *Users and Login Information*
  Are test accounts used or are users testing with live accounts?
- *Connections and relations*
  What is the connection and the relationship between the users? Are the authorizations distinct and does every user have a specific role, or are they combined in a superuser test account?
- *Permissions*
  Are the testers assigned additional permissions to perform certain actions? These would be permissions that users don't usually have but that are necessary to run the test.
- *Timeout settings*
  Are the real timeout values used or do they have to be higher or slower? Shorter periods can shorten the wait time during the test and speed up testing. A disadvantage is that it is sometimes hard to run a certain test action within the allocated time. Short timeout times are preferred to test a timeout situation, and long timeout times to test the valid flow.
- *Addresses*
  Are e-mail, technical messages or short text messages delivered to the operational addresses, or are internal test accounts used?

Metadata

Metadata is also referred to as reference data or master data. To enable a specific configuration it is sometimes necessary to predefine metadata. This can be generic data such as:

- Geo data that is used to check a zip code or a city
- User roles that are selected from a drop-down list. In this case, the list should be populated.
- Predefined statuses that an entity can have, for example, the book status "on loan" in the library system.

The physical test design contains or refers to input and output data and operational data. The configuration data and the metadata are used to set up the test environment and the smoke test.

All of the different types of test data are correlated and need to be well maintained, preferably in a test data repository.

14.2   Test Data Repository

A test data repository is a central storage location for test data. Because all types of test data (input data, expected output data, operational data, configuration data and metadata) are closely related, it is efficient to manage them as one set in a central location. This can be a tool, a test data database or an Excel spreadsheet. Managing the test data as one set reduces the risk of inconsistencies and enhances reusability.

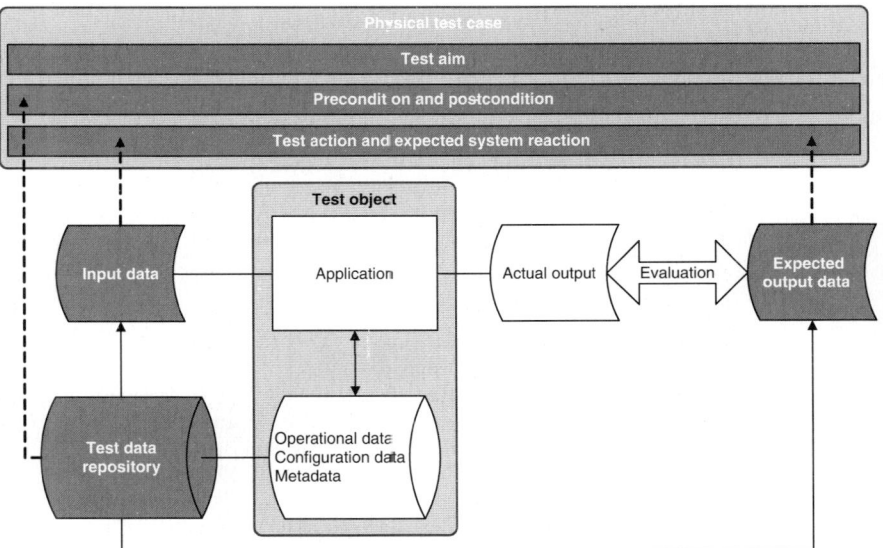

Fig. 14.1 displays the logical use of a test data repository.

The test data repository contains all of the test data. The configuration and metadata are used to install and configure the system. The live data describes the starting point of the test, which is basically the same as the precondition of the test case in the test design and can be predefined in the test object to ensure the tests run efficiently. This means that it is also included in the system database.

The input data used in the test is specified in the physical test case and entered during the test. The system reaction is compared to the expected output data, which is also specified in the test action in the physical test case.

14.3   Live Data Versus Test Data

The test data set can contain only fictitious data or of data used in the live environment . Technical test cases, such as the module tests, usually use fictitious data. During the user tests or a pilot, it is desirable to use data that closely resembles that used in the live environment.

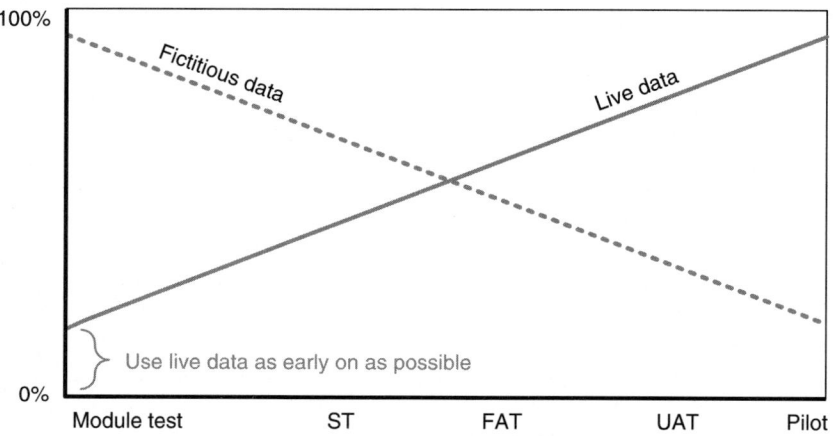

Fig. 14.2 This figure shows that the share of fictitious data decreases as the test approaches the go-live phase. Also note that it is advisable to start using live data as early on as possible. Experience shows that some errors are not found until a real data set is used.

The requirements for the test data are defined for each test level. It is wise to start using live data or data that closely resembles live data as early on in the test as possible.

Example 14.1

The system test for the Connecta project is run with fictitious test data. In order to test the authorizations properly, a separate test user is created for each user role. Because the data is fictitious, the names of the users are chosen in such a way that the testers immediately recognize which role is logged in to the application, such as FrequentFlyer or SeniorClient.

Real data is used during the UAT so that the users can get a good picture of how the application works. Certain users produce errors. Upon analysis, it seems that these users are all intermediaries that represent several organizations and several roles. Because

> the system test worked with single roles, combinations of roles
> were not tested.

In later test levels, it is preferable to work with data that closely resembles that in the live environment. There is, however, still a need for specific test data. This can, for example, be data that enables exceptional situations to be effectively tested.

Example 14.2

The maintenance organization runs an acceptance test on every maintenance release. The test uses anonymized live data and some fictitious data to test exceptional situations that would otherwise be difficult to test, such as a record in which all the fields are filled to the maximum field length. This situation does not commonly occur in practice, but the record is used to test the layout of the reports. Experience shows that changes made to the reports do not take long field names into account. The test now includes a standard check that consists of filling the fields in a report to their maximum length.

14.4   Test Data Management Strategy

Test data management consists of using the right input data and live data and maintaining the configured environment. There are a lot of ways to maintain test data. Three of the most common strategies are described below. It is also possible to combine strategies or parts of strategies.

14.4.1   Input from the Application

The easiest way of accessing the desired test data is to define it during the test. The system is configured as soon as it's available. Transactions are run during a test case to create the desired starting point.

The advantage of this strategy is that you don't have to think about the test data in advance. If necessary, the configuration of the system can be adapted during the test run to prepare it for the next test.

But this strategy also has disadvantages. Time may be saved during the creation of the test design by not preparing the test data, but because part of the preliminary work was not done, the test execution will take longer. And because the configuration of the system is not fixed and changes during the execution, some tests cannot be reproduced. This strategy does not allow for automated testing either. A tester can improvise and react to a situation; a test script can't. A further disadvantage is that the application is used to create the basic situation and configure the system. This creates a dependency on the availability of the functionality, meaning that certain tests will not run if a certain function is not available or doesn't work.

Because the test data is not thought about in advance, relationships between the different data elements are not established. This makes data management fairly simple and a test data repository superfluous.

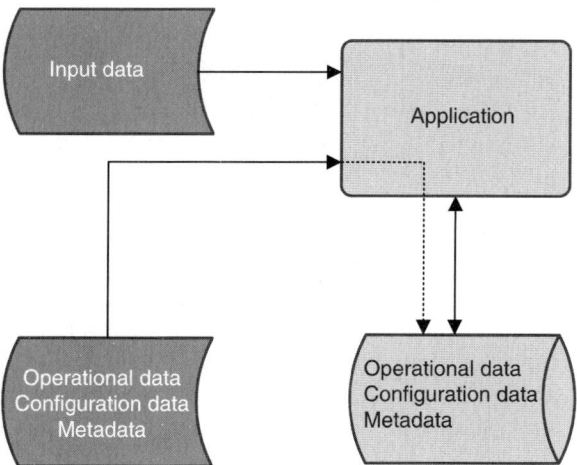

Fig. 14.3 Logical use of test data whereby all of the test data is defined during the test and entered from the application

14.4.2 *Input from the Database*

An alternative strategy is to add the test data directly to the database. This eliminates the dependency on the availability of the functionality. All of the test data can be managed centrally from the insert script that is used to populate the database. The original situation is easily restored by cleaning up the database and rerunning the insert script to repopulate the database.

The disadvantage of this strategy is that database administration (DBA) knowledge is needed to create and maintain the script. Because the test data is crucial for testing, the test team has to have DBA knowledge. Another disadvantage is that an insert script can be very complex. Errors in the script will corrupt the database, which can lead to illogical errors, such as inserts that have to be performed in the correct sequence or validations that disappear because they are usually performed at the GUI level.

There is an advantage, though: all of the test data is managed centrally. This enables a consistent set in the test data repository to be defined where configuration, starting point and input data are linked. The link between the different elements of the test data is described in Sect. 14.2.

14.4.3   Closed Loop

Test data that is added from the database (Sect. 14.4.2) generally originates from a test data repository that contains all of the test data. In the closed loop strategy, the relationship between the repository and the database is reversed. Here, a desired, already existing starting point is selected in the application database by running targeted (SQL) queries. The application database is hence feeding the repository. An advantage of the closed loop solution is that it can be based on existing (live) data. If the tests need to be rerun, it's not necessary to recreate the original situation. A query enables a new selection to be made providing the data is available in the test object database.

This works well for maintenance releases because the tests are run on an existing application and the test object database probably contains enough data. It may not work so well for new systems. In practice, the test object database usually contains enough valid data. Running tests with invalid data increases the likelihood that the query will not produce any results. In this case, the starting point will have to be recreated.

This solution can also be used in automated testing (see Sect. 14.6). The data is loaded from the test object database into the test data repository from where it is read into the test script. DBA knowledge is required to implement this solution and maintain the queries.

Step 3 – Design

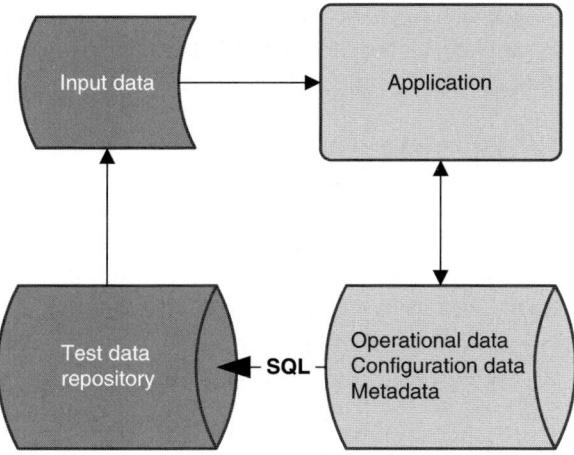

Fig. 14.4 Logical use of test data where the test data repository is populated from the application database.

14.5   Including Data in the Physical Test Design

The input data, the expected output data and the starting point are also used in the physical test case (see also the dotted lines in Fig. 14.1). The test case specifies which test data is used during the test in a number of ways. Consider using one of the following strategies:

Use Static Data in the Physical Test Case
The physical test data is included in the test design, but it is static. While designing the test, the used values can also be added to the test data repository , for example, to ensure that the system is configured in such a way that Mr. Johnson is also on the list of available users. The advantage of this strategy is that the tester knows which data he's supposed to use. The disadvantage is that using static data in the test design makes maintenance difficult and time consuming. Example of a test action: *Select a valid user (Mr. Johnson)*.

Refer to the Matching Test Data in the Physical Test Case
If the data is maintained from a test data repository, it makes sense to work in the repository because the data is immediately added to it. A reference is added to the test design so the tester knows which test data relates to which test. The below example of a test action refers to a position in an Excel spreadsheet that is used as test data repository. Referring to the data in the repository in this way considerably increases the maintainability of the test set. The test data can be changed without

having to adapt the test descriptions. A disadvantage is that the tester has to use the test design and the spreadsheet at the same time, which increases the risk of errors being made. Example of a test action: *Select a valid user (see A9).*

Use a Dynamic Link
The most advanced solution is to dynamically add the test data to the test design. Changes to the repository are immediately made in the physical test case. This is the best, but also the most complex solution. Example of a test action: *Select a valid user (Mr. Johnson).*

There is also the possibility of not including any data in the physical test design.

Do Not Include Data in the Physical Test Design
This strategy ties in well with the choice of defining test data during the test (see also Sect. 14.4.1). As a result, the test design only contains logical test data and it is up to the tester to define a physical value during the test. Although this choice gives the tester a lot of freedom during the test, it is no longer possible to reproduce the tests unless the tester notes which physical data he used during the test. On the other hand, not noting the test data saves a lot of time. An example of a test action where the test data is not defined: *Select a valid user.*

14.6   Automated Tests

For automated tests, it is important that the test data is managed centrally. If the data in the test script does not match the system configuration, the test will not finish successfully. While the manual tester can adapt his tests interactively and interpret the results, the automated test is very rigid and inflexible. For automated tests, the combination of configuration data, live data and input data needs to be 100% correct.

It is therefore advisable, if the test tool supports this, to dynamically enter the test data that is used in the tests scripts. This working method corresponds to the "dynamic link" option described in Sect. 14.5. The test data repository plays a central role in this solution as the central storage location for the test data.

Step 3 – Design

14.7   Test Data and Exploratory Testing

Up to now, we discussed test data for tests that were completely elaborated. If unscripted testing is used, as is the case with exploratory tests, what is then the importance of predefined test data?

In general, predefined input and output data is of little importance. Because the tests are created during the test run, little preparation is possible. It is, however, still important that the test environment is correctly configured. The predefined configuration data and metadata can be put to good use. The operational data describes the test's starting point. It may also be convenient to have a number of starting points ready for unscripted tests. This prevents losing a lot of time creating preconditions and enables the exploratory testers to focus on running the test charter . Optionally, the precondition can also be added to the ET test charter. See Chap. 12 for an explanation of exploratory testing and the use of test charters.

14.8   Back-up and Restore

Regardless of the strategy chosen, it is desirable and even necessary to back up the configured system and its corresponding starting points. A restore can be used to go back to the situation as it was before testing started (see also Sect. 15.6 Maintaining the Test Environment).

15 Test Environment

15.1 Introduction

Step 3 – Design

The requirements that the test environment has to meet are described in detail in the Design step (Sect. 4.4.3). We also know which tests will be run and which system configuration is needed, so we can now create the test execution environment and make sure the test can start. This chapter describes the defining and setting up of the test environment.

In a test project, planning and progress, the quality of the test design and the status of the errors require a lot of attention. It should not be forgotten that the test environment is a vital part of ensuring the test project is successful. Or in other words: the most important reason why test projects are not completed on time or do not produce the desired result is because the test environment is not good enough. Defining and setting up the test environment is often much more complex than initially thought and needs to receive sufficient attention.

It can take a lot of time to set up the test environment. If components need to be purchased, or there are problems during the set-up, the time it takes to set up and configure the environment can be substantial. This should be taken into account when the test planning is put in place.

In this chapter, the following steps are used to set up the environment:

- Determine the requirements
- Set up the environment
- Configure the system and run the smoke test

This chapter also touches on managing the test environment and the test data.

D.-J. de Grood, *TestGoal*,
DOI: 10.1007/978-3-540-78828-7_15, © Collis B.V., Leiden, The Netherlands, 2008

15.2   Determine the Requirements of the Test Environment

Before a test environment can be set up, its requirements have to be defined. The requirements depend on the tests that will be run.

The different test levels put different demands on the environment. The different quality attributes that will be tested also determine which requirements are considered more or less important. In a functional test, the focus will be on the software. When measuring the performance or running the security tests, the demands the environment puts on the hardware also have to be thought about. Imagine that performance is measured on a platform that is slower than the platform used in the live environment. The test results won't be indicative of the real performance. In order to be able to make sensible comments on the performance, the test environment has to resemble the live environment as closely as possible.

This section discusses the demands that each test level puts on the test environment. This information constitutes the background information that is needed to define the requirements of the test environment. To define the requirements of the test environment, a checklist has been created that can be used to build a bridge to the system management organization. See also Sect. 15.3.

Example 15.1

The test coordinator wants to run performance tests. On the checklist, he indicates that he needs a fast network connection between the client and the database server. He and the system administrator determine what "fast" means and which network components (hubs, network cards) will be included in the chain. They check whether the components meet the defined performance requirements. System management uses the checklist and the information from this session to set up the test environment.

15.2.1   *Module Tests and Module Integration Tests*

Module tests, which are sometimes referred to as unit tests (UT), are "white-box" tests. White-box tests are tests that provide insight into the code and where the internal workings of the system are known. The test environment is usually the same environment as the one the system is being developed on. A development environment is equipped with, among other things, a compiler, an SQL client, etc.

A unit/model test tool is used to run unit tests; in Java environments, for example, JUnit is used. Because the modules are tested at a point in time when not all of the modules are available, missing modules are simulated with stubs and drivers. It is fairly common for a developer to run the module tests rather than a tester. The deployment cycle is short, which means that there is not much time between finding an error and fixing it. The developer tests, finds an error, fixes the error, and runs another test on a new version of the module.

Developers will often use their own test data, which often consists of technical data that resembles the live data used in functional tests. For module integration tests, the test data will often be stored centrally so that the process becomes more efficient by reusing test data. This central storage location is the test data repository (see also Chap. 14 Test Data).

Requisites for module tests:

- Development tool
- SQL client or database administration tool
- Unit test tool
- Stubs
- Drivers
- Test data repository
- (Technical) test data
- Bug tracking system
- Version management tool

15.2.2   System Tests

System tests (ST) are black-box tests; they provide no insight into the code and the system is tested from the outside. A large part of the tests are run using already developed system interfaces. The system test is often run at a point in time when not all of the system components are available, which is why they are simulated by stubs and drivers. For example, a Web application can be tested without actually connecting it to the Internet or to the network because it is run on a simulated server.

The system test does not look into the system, although there is a need to view the system logs and message queues. The testers perform actions directly on the database, which requires the testers to have read

permissions on all of the system elements. In consultation with the environment manager, it can be convenient to give one or more of the team's testers write permissions so that he can optimally control the environment. The testers who are granted write permissions need to have enough knowledge and experience so they can assess the consequences of their changes.

System tests are no longer run by developers, which is why the deployment cycle of the system test will be somewhat longer than that of the module test. It is desirable that showstoppers are solved quickly and the software released to the testers.

Tests are run using test data, which is stored in a test data repository (see also Chap. 14 Test Data) that is shared by the testers. If a tool is used to design or run the tests, the tool should also be available.

Requisites:

- Stubs
- Drivers
- Simulators
- SQL client or database administration tool
- Monitoring tools
- Read permissions for all of the log files, databases and system elements
- Write permissions for all of the databases and system components
- Test tool (test automation and/or result logging)
- Bug tracking system
- Test data
- Test data repository

15.2.3   *Functional Acceptation Tests*

Just as for system testing, functional acceptance tests (FAT) are mainly run using the developed system interfaces. For the FAT, the system is expected to be developed to such a degree that stubs and drivers are no longer needed. It may, however, still be necessary to simulate external systems.

In this phase, the system can be expected to work fairly well. The errors found during the system test have helped improve the functioning of the system. It's no longer necessary to examine all kinds of system logs. System logs contain errors that are not intended for users and are hence not displayed in the user interface.

It is convenient if the testers have read permissions on the database and the primary system logs. It is not necessary (and even not desirable) that the testers change the configuration or the database. Experience shows that changes that are made without the knowledge of the administrator are often forgotten. The result is that an error will occur again when a new version is installed. In the best case, the tester will remember that the error is easy to fix and will fix it again; in the worst case, a lot of unnecessary time will be spent analyzing the error. Make sure you don't lose control of the configuration of the test environment. Changes should go through the manager and be checked to ensure they are implemented in the live environment.

Compared to the system test, the deployment cycle can be longer (the new version is available the next day). The FAT is a controlled test phase, the smoke test has proven that the system is working reasonably well and is ready for testing.

It's obvious that the test designs, the test tools and the error logs have to be available. The tests will use test data, for example, for syntax testing, but also to test live data. This data may also have to be anonymized.

Requisites:

- Simulators
- SQL client or database administration tool
- Read permissions for system log and databases
- Test tool (test automation and/or result logs)
- Bug tracking system
- Test data repository
- Test data or anonymized live data

15.2.4 User Acceptance Tests

The user acceptance test (UAT) is run together with users or user representatives. The tests consist of realistic scenarios that are created together with the users. The system should be stable enough to support the primary business processes. External systems can be simulated. The users do not normally need to have technical knowledge to run the tests.

As the UAT is usually run on a stable system, the deployment cycle does not necessarily have to be short. It is not expected that many showstoppers will occur. If major flaws are found during the smoke test, the UAT will be suspended.

The tests will use either live data or representative test data. The test data may consist of anonymized live data.

Requisites:

- Simulators
- Bug tracking system
- Test data repository
- Test data or anonymized live data

15.2.5   *Production Acceptance Tests*

The production acceptance test (PAT) does not focus on the functionality of the software, but on the maintenance aspects. The PAT consists of an elaborate checklist and the testing of maintenance. Filling out the checklist helps ensure that systems that are entering the maintenance phase were tested, the system documentation delivered, agreements made about support, etc.

Among other things, maintenance testing consists of

- Back ups
- Restores
- Installing a new version
- Support procedures
- Fallback scenarios and implementation

The things that are needed are the same as the things the maintenance organization needs for the live environment. See also Sect. 4.2.

15.2.6   *Chain Tests*

In a chain test, two or more systems are connected. In previous test levels, simulators replaced the missing components. In the meantime, those components will have been completed, which means that they can be used instead of the simulators. Experience shows that a lot of errors are found during this round of integration testing. The chain tester should have access to the database and system logs so he can analyze what is causing an error. Technically speaking, the chain test is very much like the system test, but in terms of organization it is very different.

Chain tests are often very complex because of the many parties and systems that are involved in the test. The system that is being tested has connections to other systems, some of which can also be connected to other systems. Because the chain can become very long, the chain is often only set up with the systems that are directly connected to the system that is being tested.

A lot of the complexity is due to the fact that all of the systems in the chain need to use the same test data. If one system uses a customer number, all of the systems need to use the same customer number. Needless to say, the data that is related to the customer number also has to be correct. Because existing test environments are often used for the chain test, it is very difficult to harmonize the test data across the systems.

If a chain test is planned, agree with the chain partners on the test data at an early stage. For example, you can agree to use the same customer data, or to use fixed ranges for test order numbers.

Step 3 – Design

15.2.7   Pilot

The pilot is a way of running the live environment in a safe environment. Here, safe means that the impact of possible problems is minimized, for example, by running a shadow of the existing system. Should something go wrong, you can fall back on the existing system. For new systems (or services), the pilot is also used to run tests with a small group of users. This has the advantage that only a small group of users will be affected should unforeseen problems occur.

In a pilot, the rollback procedure needs to be given enough attention especially if the pilot environment is connected to chain partners. If the live environment needs to be restored to its original configuration, transactions may need to be rolled back on other systems in the chain. The pilot can be run in the live environment as well as in the test environment. In the latter case, the requirements for the test environment are the same as those for the live environment.

15.2.8   Performance Tests

Performance tests consist of doing load, stress, reliability, and concurrency measurements (see also Sect. 4.7). This results in requirements for the test environment that are different from those for functional tests. For performance tests, the hardware and the infrastructure should be representative of the live environment. This means that the test environment has to be the same as the live environment. It's sometimes possible to convert the results to the performance that is expected when the test object really is running in the live environment. Performance testing also includes testing the network connections between two systems in the chain.

The test data used for performance tests should resemble the live data as closely as possible. Queries take longer on a populated database than on an empty database, and the time it takes to process messages (for example, the time needed to parse an XML message) that are sent or received by the system will vary depending on the size of the message. The messages that are used in the performance test have to be representative of the ones produced in the live environment.

Performance can be analyzed using the time a process step takes. The system resources that are used also provide important information. The development environments often provide their own analysis and monitoring possibilities. Logging in to the server also produces information. If the standard information and tools are not sufficient, additional measuring points will have to be added to the code. These measuring points are called hooks.

To measure the performance, it is desirable to put a certain load on the system for a longer period of time. This load simulates the behavior of several users performing transactions on the system. Tools and simulators used to do this should be available and are placed on a separate system in order to reduce their influence on the performance of the test object.

Reliability tests often have a longer runtime and are run at night. Stress tests are often run outside office hours because the heavy load used in these tests puts a heavy load on the network and the servers. It is important that the test can run uninterrupted, which is why processes that could influence the test need to be taken into account.

In many organizations, processes are run at night. For example, the server is restarted, batch jobs are run and back-ups made. Open connections are sometimes automatically closed. Establish which processes are running and how their influence can be minimized. If this is not possible, some of the problems can be solved by using a separate network segment.

Performance measurements are often analyzed with special tools. The results are presented in graphics.

Requisites:

- Hardware that is representative of the live environment
- Same infrastructure (for example, a network connection) as the live environment
- Simulators
- Load generators
- A populated database with representative test data consisting of anonymized live data
- Test data repository
- A separate system on which simulators, generators and measuring tools are run
- Performance test tool
- Network load measuring tool (analyzes the behavior of the application and explains bottlenecks)
- Permissions to access logs and resource utilization on the server
- Bug tracking system
- Result analysis tool
- Graphics program

15.2.9   Security Tests

Security tests can be run at a number of testing levels. The tests focus on different parts of the system, which consists of the modules, the application, the procedures and the infrastructure in which the system operates.

In module security testing, components or objects are tested for security flaws. It is usually quite simple to extend unit test environments with test cases for security testing, but specialized software that runs these tests is also available. In addition to unit testing, tests are also run on the coding standards or security guidelines the code is based on. In this case, the requirements for the test environment are the same as for the unit test.

Application security tests test the whole application. Just as with unit testing, the environment that can be used for application security tests is the same as the one used for system tests.

Network security tests map out flaws by running a test from the network, such as a hacker test, which checks whether unauthorized individuals can access the system. It is important that the infrastructure is the same as in the live environment. It will not be possible to ensure that the security in the live environment is good if the test environment uses a different firewall or configuration. In practice, this level of testing is often run on the live environment with the necessary precautions.

Requisites:

- See module and module integration tests (15.2.1)
- See system tests (15.2.2)
- See pilot (15.2.7)

15.2.10 Training Purposes

If a test environment is designed for training purposes it will closely resemble the environment used for the UAT. The stability and the performance of the environment need to be in order. Experience shows that the training manuals are often based on previously defined starting points. This enables the participants to run through some of the scenarios very quickly. The starting points should be available in the system, which puts demands on the test data. It must also be possible to use the original data set in a subsequent course. See also Sect. 15.6.3.

During the training, simulators can be used to replace external systems or to create error situations. The participants should not need extensive technical knowledge to follow the training.

As opposed to the UAT, the purpose of the training is not to find errors, which is why a bug tracking system is not necessary. It is however, helpful if the trainer is able to log errors.

Requisites:

- Simulators
- Training data or (anonymized) live data
- Bug tracking system

15.3   Test Environment Requirements Checklist

This checklist can be used to identify the requirements of the test environment.

The tester who is setting up the test environment fills out the checklist and discusses it with the test coordinator. Together they check whether the requirements for the test environment support the test strategy.

The purpose of the checklist is to help build a bridge to the system management organization. While discussing the list, the system administrator gets a picture of the requisites and fills out the rest of the list. This requirements list is used to decide how the environment will be set up and which hardware needs to be purchased.

Requirement	Option
1. Type of environment	☐ Development ☐ Test ☐ Acceptance ☐ Live
2. Which test levels does the environment have to support	☐ Module tests ☐ System tests ☐ Functional acceptance tests ☐ Production acceptance tests ☐ Chain tests ☐ Pilot ☐ Performance tests ☐ Security tests ☐ Training ☐ Other …
3. Configuration check	☐ Complete (it is not possible to change anything in the software versions, in the configuration or in versions of supporting packages like DBMS, server software etc.) ☐ Partial (releases, patches and upgrades of supported packages are controlled, but the configuration can be changed without notice.) ☐ None (everybody can change everything, there is no control)
4. Black box or white-box testing	☐ White box, check whether development tools are needed. ☐ Black box
5. System components	☐ Standalone ☐ Client, quantity: … ☐ Server (for example, for database, Apache, active directory, etc.), quantity:… ☐ Network components (hub, router, switch, UTP cables, etc.) ☐ Generic platform ☐ Other …

Step 3 – Design

Requirement	Option
6. Resemblance with live environment	☐ Hardware same as live environment ☐ Infrastructure (for example, network connection and redundancy) same as live environment ☐ Software configuration same as live environment ☐ Software not same as live environment, there is a special test build with additional hooks and checks
7. Performance	☐ Network performance: High/Standard/N/A ☐ Server performance: High/Standard/N/A ☐ Client performance: High/Standard/N/A ☐ Peripheral equipment (hub, router): High/Standard/N/A
8. Concurrent users	☐ Number of testers simultaneously working on the system: …
9. Interfaces with external systems	☐ Yes … ☐ Simulated … ☐ No
10. Connection with external systems	☐ Through a network ☐ Through the Internet ☐ Through message exchange ☐ Through batch (e-mail, data storage exchange) ☐ N/A
11. Software security	☐ None ☐ Log in ☐ Encryption ☐ Authentication ☐ Firewalls ☐ Other …
12. Physical security	☐ None ☐ Test environment is in a separate room only accessible to staff ☐ All components and all data are in a separate room only accessible to authorized staff
13. Accessibility	☐ The test environment is accessed from the standard workplace ☐ The test environment is accessed from a separate PC, additional requirements are: … ☐ The test environment is part of the standard business network ☐ The test environment is part of an standalone development/test network
14. Protection	☐ Permissions to view or edit system logs ☐ Permissions to view or edit queue ☐ Permissions to view or edit database ☐ Permissions to view or change system configuration ☐ Permissions to view or change system users ☐ Permissions to use the system as a user ☐ Permissions to work as an administrator or user in interfacing systems
15. Tools	☐ Test data repository ☐ Bug tracking system ☐ Development tool

Requirement	Option
	☐ Unit test tool ☐ SQL client or database administration tool ☐ Drivers ☐ Stubs ☐ Simulators ☐ Monitoring tools ☐ Test tool (create load, measure performance) ☐ Result analysis tool ☐ Test tool (test automation and/or result logging) ☐ Separate machines to run simulators and test tools ☐ XML editor ☐ Text editor ☐ Message editor/generator ☐ Data analysis tool ☐ Version management tool ☐ Other...
16. License policy, additional tools	☐ Only freeware (no license costs) ☐ Only limited license costs (€ ... max.) ☐ Only after extensive tool selection, no maximum, but business case required
17. Stability and uptime	☐ Automatic restart ☐ Yes/No ☐ Automatic deployments ☐ Yes/No ☐ Automatic processes ☐ Yes/No ☐ Automatic closing down of connections ☐ Yes/No
18. Speed of deployment cycle	☐ Within 1 hour ☐ Within 1 day ☐ Other...
19. Test data	☐ Technical test data ☐ Logical test data ☐ Training data ☐ Live data (anonymized) ☐ Live data (not anonymized) ☐ Live data (migrated)
20. Size of the test data set	☐ Small size < ... MB ☐ Medium size < ... GB ☐ Large size < ... GB
21. Sharing information	☐ Own data, own system ☐ Same test data, but own data set to work with ☐ Same test data, all testers work with the same database
22. Test data, maintenance	☐ Test data is defined by the tester, no fixed test data ☐ Test data has been defined and is maintained centrally. Test data repository ☐ Test data is rolled back to the starting point after each test
23. Test data, starting points and configuration	☐ Are always implemented from the application ☐ Are initially implemented from the application and are rolled back to the starting point after each test.

Step 3 – Design

Requirement	Option
	□ Are entered from the database using SQL □ Other …
24. System needs to work with different configurations	□ Works with following versions of MS Windows … □ Works with following versions of Web browsers … □ Works with following versions of DBMS … □ Works with another OS… □ Works with other components …
25. System availability, sharing with other projects	□ The environment has been assigned to this project only □ The environment is shared, but software installations are independent, switching has no impact □ The environment is shared, changing software components or configuration is necessary and switching has impact
26. Compatibility	□ Test environment must be interchangeable with other environments (Development, test, Acceptance, Live) □ Test environment should be interchangeable with other environments (e. g. Acceptance 1, Acceptance 2, etc.)
27. Manual or automated test execution	□ Manual □ Automated
28. Test environment is installed by	□ Test team □ Development department □ Maintenance department
29. Additional tools are installed by	□ Test team □ Development department □ Maintenance department
30. Maintenance and support	□ Own maintenance □ Maintenance by development department □ Maintenance by maintenance department
31. Knowledge and skills of management team	□ System knowledge □ Network knowledge □ Database knowledge □ SQL □ Tool knowledge □ Other …

15.4   Setting up the Test Environment

The completed checklist is used to set up the test environment. The maintenance organization (or other parties responsible for the test environment) can use the requirements to acquire the components and implement the test environment.

The set-up consists of the following steps:

1. Acquire the infrastructure
2. Install the infrastructure (for example, DBMS)

3. Install the metadata (see also Chap. 14, Test data)
4. Install the test object (pre-release)
5. Test the test environment
6. Install the test object (release that is to be tested)

This list distinguishes between installing the pre-release and installing the release of the test object. The distinction is made in order to test the test environment and the installation procedure in advance to reduce the risk of having to use valuable testing time to install the system when the test object becomes available. At this point, you want to start testing, not spend valuable time on the test environment. In some cases, the release that is to be tested is immediately available and the environment is tested at the same time the smoke test is run (see also Sect. 15.5.2).

15.5   Configuration and Smoke Test

15.5.1   *Configuring the Test Environment*

When the test environment has been set up and the test object has become available, the environment and the test object can be configured. This activity can overlap with the previous activity, since it is necessary to configure the system during set-up. Configuration is a necessity if it is not possible to test the test environment without partially configuring the test object. Configuration using a pre-release is desirable, because it prevents surprises arising. Only the changes will have to be implemented in the final release once it becomes available. This situation also occurs when the release is being used for a retest or for a regression test.

Configuration data is used to configure the test environment. This data is part of the test data ; see also Chap. 14 Test data.

15.5.2   *Smoke Test*

The purpose of the smoke test is to ensure that the quality of the system is good enough to start testing. The smoke test is run before testing starts and as soon as the version of the test object to be tested has been released.

The most important reason to run the smoke test is to save testing time. Experience shows that it is very likely that the delivered systems will not work as desired, meaning that it will not be possible to run a reliable test on them. If the delivered system does not work as desired, it will

Step 3 – Design

have to be flagged before the test starts so measures can be taken. The smoke test is described in Chap. 17.

The smoke test is run on a "tuned" system, meaning that the system has to be configured before the smoke test can be run. In practice, configuration and smoke test are not always distinct activities. The system is used to check the changes made to the configuration. To run a smoke test on the system, the test environment needs to be configured. Possible errors can be caused by bugs in the software, but also by configuration errors. The lack of distinction between the configuration and the smoke test doesn't matter as long as the purpose of the smoke test is not forgotten.

15.6   Maintaining the Test Environment

Maintaining the test environment is at least as important as setting it up. If the test environment works properly during the first test run, it will probably work properly during the retests and the regression tests. To ensure the test environment remains a reliable factor throughout the test project, its maintenance needs to be well organized. If the test environment is maintained by an in-house maintenance organization, everything discussed in this chapter will most likely have been dealt with. If the test environment is not maintained by an in-house maintenance organization,  the test coordinator will have to set up the maintenance himself. If there is a standard procedure that can be reused, it is advisable to use it because it increases efficiency and makes the transition from project to maintenance easier. The below activities should be well organized.

15.6.1   Configuration Management

Configuration management is the process that ensures that the structure of the test environment is known and controlled. Changes to the test environment often have an impact on the test results, which is why it is important to know which changes are implemented when. The version and configuration of all of the relevant components are documented. These components are also referred to as configuration items [ITSMF, 2000].

In your project, make sure the following information is maintained:

- Version of the test object
- Version of the supporting software packages

- Version of all of the patches that were applied
- Version of test data repository
- Version control (what has been changed when)

15.6.2 Release Management

It is important that the changes made to the test environment are controlled. Release management ensures that changes are implemented at previously agreed times and that the implemented changes are documented.

Releases are often released in consultation with the test coordinator, the project leader and the administrator. The test coordinator indicates the ideal time for a new release. The project leader knows how the team is progressing and can indicate how much time he needs to create a new version of the test object. The operator is responsible for the release and plans the activities. A release is not only accompanied by a new version of the test object, but also by release notes and an installation or deployment guide.

Release Notes
Release notes describe the characteristics of the deliverable. Release notes include:

- The version of the release
- Changes in relation to the previous release
- The errors that were fixed
- The implemented change requests (RfCs)
- The known errors that can impact the test process

Installation or Deployment Guide
The installation or deployment guide includes instructions for the administrator. These instructions can have the form of a checklist and indicate which installations still need to be done. For big systems, the deployment can be very complex. The sequence in which the installations, upgrades, patches and controls need to be run should be specified. The installation or deployment guide is often written by the maintenance organization together with the party that built the system. A good installation or deployment guide ensures the installation runs smoothly and prevents problems occurring during the test run and in the live environment.

Step 3 – Design

15.6.3   *Back-up and Restore*

It goes without saying that the test environment should be backed up before an installation and at regular intervals. Should an installation go wrong or the environment become corrupt, the back-up can be used to restore the test environment.

Even if errors don't occur, you may want to restore the test environment to its starting point, for example, to rerun a test or run a regression test. Restoring a back-up means that a test can be rerun with the same test data. This is relevant, for example, if a test requires creating unique records. If the data is dynamically entered in the test scripts (see also Sect. 14.4 Test Data Management Strategy), it may be quicker to use different test data than to do a restore.

When restoring a database from back-up, remember that the database structure may have changed in the new version of the test object. Restrictions and stored procedures may also have changed. After the database has been restored from the back-up, all the patches will have to be applied. Note that the back-up is restored in agreement with the test coordinator (I want to go back to that situation), the maintenance organization (we will be able to do it at this time) and the party building the system (these are the changes we made since the previous version).

Step 4 – Set up

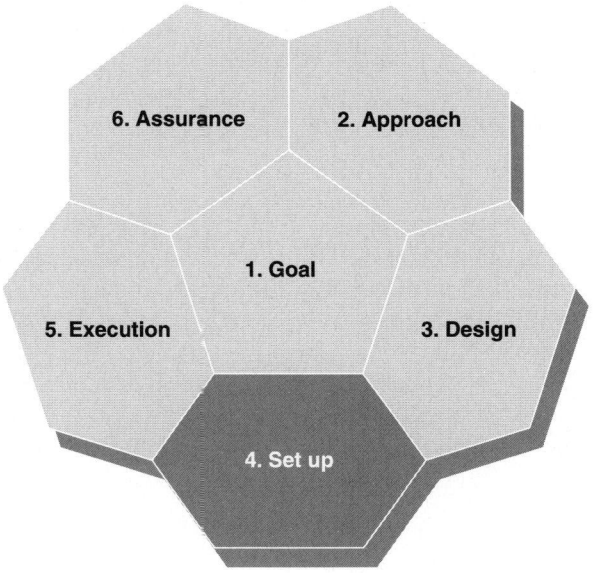

During the Set-up phase, tests are scripted and the test environment created. Creation means acquiring and developing the items that are specified in the requirements of the test environment. After a successful smoke test, the created environment is finalized, meaning that the test environment now supports the execution of the physical test design and the scripts.

Set-up consists of the following activities and products:

Activity	Product
Set up the test environment Configure the system	Test environment
Test automation	Scripts for automated testing
Smoke test	Smoke test errors Smoke test report

16   Test Automation

16.1   Introduction

A lot of organizations wonder whether it's worth switching to automated testing. The concept presented by commercial tool suppliers is
often very tempting: test while you sleep [Siteur, 2005]. Unfortunately,
reality is different. It may not be worth automating all of the test cases.

There are a number of reasons why manual testing will continue to be
necessary; the automated testing of requirements with a high change-
ability, uncertainty or complexity is very expensive. Moreover, automated tests need to be prepared and maintained, just like manual tests.
Automated testing enables a lot of tests to be run in a short period of
time. Nonetheless, the test results still have to be checked by a test expert, a process that is not much different than that for manual testing. In
both cases, tests are prepared and run and in both cases the errors are
logged, examined, fixed and retested. But that doesn't mean that everything has been said and done. This chapter discusses automated testing
in depth because there are a lot of situations in which it makes sense to
run automated tests. This chapter takes a closer look at situations in
which test automation is a technical necessity as well as at the ways in
which test scripts are created. A lot of books have been written about
test automation [Fewster et al, 1999], [Broekman et al, 2001], [Dustin et
al, 2005], which is why this chapter will focus on creating a conceptual
framework that the test expert can use to discuss the basic principles
with his organization.

D.-J. de Grood, *TestGoal*,
DOI: 10.1007/978-3-540-78828-7_16, © Collis B.V., Leiden, The Netherlands, 2008

16.2   What is Test Automation?

When people hear "test automation" they immediately think of the automated execution of tests. The picture they have is that tests are run at the press of a button or the click of a mouse, which makes them cheaper and easier to run than manual tests. Unfortunately, things are not that simple. On the one hand, this picture is far too positive, and on the other hand there are many more forms of test automation . In Test-Goal, test automation is defined as follows:

> Test automation is the use of test tools that support the test process and help the tester find errors.

If, for example, errors are logged in an error log, we call that test automation. Test automation is also the use of simulators to replace missing system components. There are different types of test tools that can be used to automate tests. We distinguish between

- Dynamic tools
- Static tools
- Supporting tools

16.3   Dynamic Test Tools

Dynamic test tools are used for dynamic tests, which are tests in which the system is actually used. Dynamic test tools are used for one or more of the following reasons:

16.3.1   *Additional Testing Possibilities*

Some tests are almost impossible to run manually. Dynamic test tools can make a lot of tests easier to run, but they are an absolute necessity when it comes to testing an object that does not have a user interface, like a smart card. In this case, a tool is needed that sends commands to the card and looks at the response the card returns. Another example is module tests, which are difficult to do manually. Because module tests focus on small parts of the code, stubs, drivers and automated scripts are needed to check if the modules technically meet the system requirements.

A dynamic test tool is not only used to generate good behavior, but also to create error situations (how will the test object react to wrong input). The correctness, specific content, consistency and syntax of the test object's response can be checked automatically.

16.3.2   Time Saving

There are tests that take up too much time when run manually, such as security tests, which require checking a lot of options to ensure there are no leaks in the security protocol. Or reliability or endurance tests, in which tests are repeated many times to determine whether the system is stable. Using simulators to automate tests can also save a lot of time, for example, if manually simulating a peripheral system is labor-intensive and prone to error. In this case, simulating the peripheral system can significantly increase the efficiency of the test run.

16.3.3   Log files

To record information that is (virtually) impossible to obtain manually. An example of this is performance testing, where it's very difficult to measure times in real time if the steps succeed each other very rapidly. Tools can be used to measure and record the performance accurately.

```
13-2-2006 16:01:07  ,1-12,POST http://www.empe3.nl/index.php?ma
13-2-2006 16:01:08  ,1-12,GET http://www.empe3.nl/index.php?mai
13-2-2006 16:01:08  ,1-11,GET http://www.empe3.nl/index.php?mai
13-2-2006 16:01:09  ,1-11,POST http://www.empe3.nl/index.php?ma
13-2-2006 16:01:10  ,1-11,GET http://www.empe3.nl/index.php?mai
13-2-2006 16:01:10  ,1-10,GET http://www.empe3.nl/index.php?mai
13-2-2006 16:01:10  ,1-12,GET http://www.empe3.nl/index.php?mai
```

Fig. 16.1 Fragment of a system log. On the left, the timestamps indicate when the http requests were executed.

16.3.4   Comparing Results

The ability to compare the observed results with the expected results objectively. The comparison of the results can be included in the test script and can take place during the automated test to establish the conclusion of the test. If the observed result is not the same as the expectation defined in the script, the test will receive the status "failed" (see also Fig. 16.2).

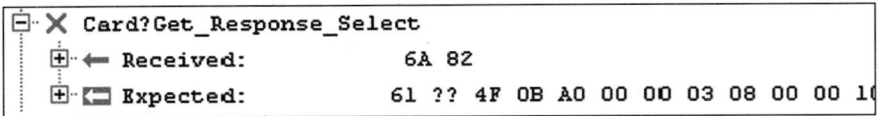

Fig. 16.2 Fragment of a test log of an automated protocol test. The log shows that the test was not successful.

Results can also be compared after the test, for example, the comparison of big messages that the system sends. Manual comparison is very difficult and a lot less objective than automated comparison. Microsoft Word can, for example, be used to compare the two messages to each other. This suddenly promotes the text editor to a testing tool.

16.3.5   *Extensive Repeatability*

To execute the same test again and again in the same way, possibly for various versions of the product. This is important for, among other things, conformity tests (the tests have to be executed in the same objective way for different suppliers) and regression tests (the same test has to be run on a new version of the product).

The ideal dynamic test tool supports the tester in all the above-mentioned areas and provides a number of functionalities for each of the previously mentioned reasons. The additional functionality is displayed in Fig. 16.3. The figure provides an overview of the desired possibilities.

Record and Playback
The ability to record the actions a user is performing on the test object, store them in a script, and play them back.

Scripts
The program code that runs the tests. This program is coded manually or is automatically generated by the record and playback function.

Input Data and Expected Output Data
The ability to save the input data and the expected output data separately and to use it to run the scripts.

Result Logs
Comprehensive logs that include the results of the test run as well as of the performance measurements of the test object.

Comparator
Compares the measured results to the expected results.

Test scenario
The ability to merge available tests or scripts into test scenarios.

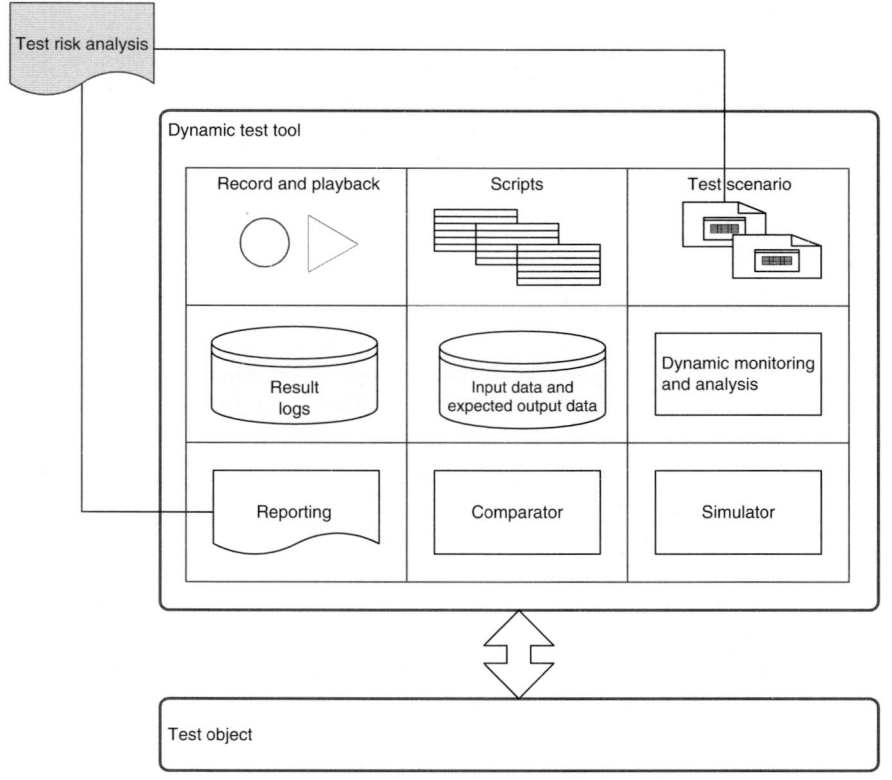

Fig. 16.3 The ideal test tool supports a lot of functions

Dynamic Monitoring and Analysis
The ability to monitor the test object and analyze the cause of a problem (for example, resource allocation, network monitoring, memory leaks). Other possibilities are test coverage tools that do more than monitor the test object. The test tool ads code to the object to trace which code is affected during the test execution.

Reporting
Present the results and output of the comparator in a report that ties in well with the test risk analysis.

Simulator
The test tool communicates with the test object and simulates a user or another system.

There are not many dynamic test tools that offer all of the above-mentioned features, which is why it is often necessary to use a combination of dynamic test tools (see also Sect. 16.8)

16.4   Static Test Tools

Static test tools are used in static tests: the tests that do not require the program to be started. Static test tools can focus on the software or the documentation.

A few examples of static test tools:

- Tools to measure the complexity and the structure of the code (for example, the average size of a function, the number of nested IF-THEN loops, etc.)
- Tools to check the correctness of the code (for example, to detect unused routines)
- A spell checker for the system documentation

16.5   Supporting Tools

In addition to the tools that help run the tests or analyze data, there are also test tools that support the test process in general, such as:

- Error logging tools (bug tracking)
- Planning tools
- Tools to create test designs
- Tools that can be used to report the results of manual or automated test runs (dashboard)
- Configuration management tools

16.6   Test Automation: Yes/No

16.6.1   Business Case

The previous section revealed that automation is not always applicable. In some cases, the nature of the tests or of the test object will dictate the use of tools. In other cases, there will be a choice, and in most cases the decision will be based on a cost-benefit analysis.

The costs of an automated test include the cost of selecting, acquiring and implementing the tool. Moreover, it takes time to learn how to use the tool; unproductive time that costs money. After the tool has been implemented, the costs mainly consist of:

- License fees
- The cost of developing test scripts and/or tools
- The cost of maintaining test scripts and/or tools

These are costs that are not incurred by manual testing. To build a favorable business case, the costs are generally cut out of the test execution. This doesn't pose a problem if the tests are frequently repeated and the test set requires little maintenance. A rule of thumb is that if the automated tests are run less than five times, the investment is not earned back. This rule of thumb applies to all automated tests, but not all tests need to be automated. For organizations that are starting to automate their tests, the best advice is to automate only the tests that have to be run a lot of times or if automation saves a lot of time.

The maintenance costs of an automated test set depend on the stability of the test set and the ease with which the tests can be modified. For conformity and interoperability tests (see also Sect. 4.6) the test base is often very stabile and the tests can be reused without a lot of maintenance.

In a situation where the test base is constantly changing, more maintenance will be needed. Maintenance costs money and takes time. However, without good maintenance, the testware will quickly become useless. This can be prevented by setting up the implementation in a flexible way and by ensuring that changes are easy to implement. Although the initial development costs are often high, the maintenance costs will decrease, as is depicted in Fig. 16.4.

The graphic shows that an optimum can be reached between development costs and maintenance costs. The precise form of the curves, and

Step 4 – Set up

hence of the location of the optimum, depends on many factors, such as the test team's experience with scripting and with the tools that are used. If the test team is very experienced, the development costs will be lower than if the team is not very experienced. The curve for the development costs will be less steep and the optimum will shift to the right. Depending on the stability of the test object and the test base, the maintenance costs will also vary. In a very stable test base, the maintenance costs may already be low, meaning that implementing a more flexible automation will not result in higher savings. The curve for the maintenance costs will start low and get flatter, and the optimum will shift to the left.

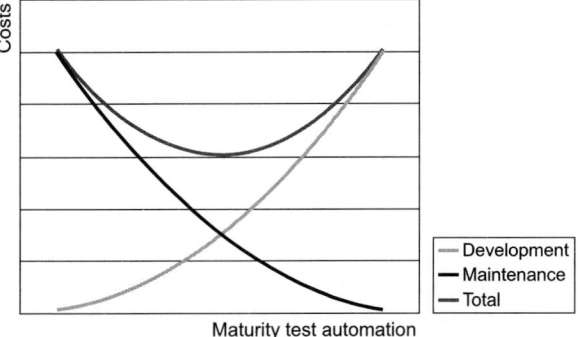

Fig. 16.4 As the flexibility of the implementation increases, the maintenance costs decrease and the development costs increase.

In addition to reducing costs, there are other reasons why test automation can be worthwhile:

Technical Necessity
Some tests simply can't be run manually. A few examples were discussed in this chapter.

Time to Market
The test run is usually in the planning's critical path. When running a quick scan or a regression test in a maintenance environment, it is often desirable that the tester give a release advice quickly; the time to market is not allowed to suffer from the time lost running the test. Test automation provides a solution. If the automated tests prove themselves during the implementation, they can soon be repeated.

Reproducibility
There is always a certain amount of variability when tests are run manually. Test automation provides a solution if it is important that tests are run in exactly the same way every time.

Controllability

Some organizations have to prove that enough testing was done. This can be due to legal regulations or a fear of damage claims. Having test scripts that accurately show how the tests were carried out and what exactly was tested, for example, as part of a certification process, can be the decisive argument.

Motivation of the Test Team

Test fatigue will set in if tests are repeated over and over again. The tester may lose his objectivity and motivation. This benefits neither the employment relationship nor the quality of the tests. Test automation can ensure the test expert will not spend valuable time on repetitive work and will instead be able to focus more on the anticipated goal.

16.6.2 Making a Well-Informed Decision

Test automation brings software development into the testing domain. What applies to developing the test object also applies to test automation. Paying attention to the requirements for the test tools and test scripts, good version management and expertise in test automation increase the likelihood of success.

If the functionality of the test object is not clear, a lot of energy can be put into writing nice scripts and simulators, but it will still be hard to develop good tests. If the functionality changes significantly with every release, more time will be spent on maintenance than using the testware. The stability of the test object is hence a precondition. Another precondition is that there is a stable test environment with which the scripts and tools for development and testing are integrated. Don't forget that testware also has to be tested [Zambelich]. The test team needs to have the expertise and the testing time will have to be scheduled.

16.7 Developing Test Scripts

A test tool is needed to run automated tests. But a test tool on its own doesn't do a lot; instructions are needed to tell the tool which test actions and controls need to be carried out. These instructions are provided in the form of test scripts, which are test cases that have been converted into code. A test script is created after the physical test case has been created, or the test script can be the physical test case. A number of strategies, which vary in maturity and complexity, can be used to develop scripts. The boundary between a physical test case and the script decreases as the test scripts become more mature.

As a reference, the manual test is first displayed in Fig. 16.5. The tester creates a logical test design (LTD) and elaborates it into a physical test design (PTD). The physical test design is his plan of action during the test.

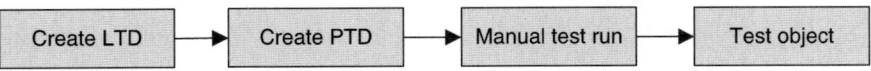

Fig. 16.5 Manual test run

When the tests are automated, a number of things in the above figure change. The following sections explain how test scripts are created by a record and playback tool and by programming them.

16.7.1 *Record and Playback*

One form of scripting is record and playback. The test tool records the tester's actions in a script, which can be played back and changed if necessary. A user interface is needed to do this. This form of scripting is often used in a system with a graphical user interface (GUI). Record and playback can be done in the following ways:

Linear Scripting
Linear scripting consists of generating scripts by recording the tester's actions during a manual test run using a record and playback tool. The test actions must be predefined in a physical test case that is completely separate from the script.

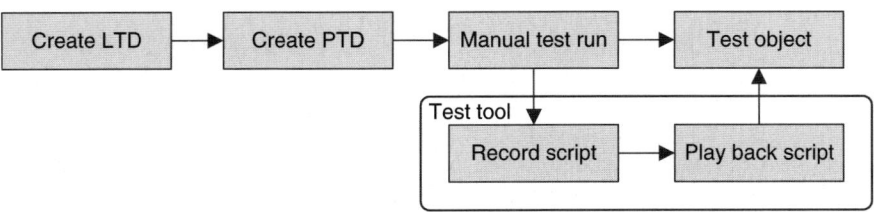

Fig. 16.6 Record and playback scripting

The advantage of linear scripting is that, because the scripts are recorded, it doesn't take much additional time to create them. The disadvantage is that the scripts are not very flexible and control over the content is limited since all of the tester's actions, including the errors, are recorded. Test data is embedded in the script making it difficult to change it. The reusability of the script rapidly decreases when the system or the data changes during the test run.

Structured and Data-Driven Scripting

Structuring linearly generated scripts increases the control over the flow of the scripts. Using conditions and statements like "if-then-else," "for" and "while," makes the scripts more flexible. Shared functionality can be reused and certain tasks can be executed iteratively. The physical test case is still separated from the script; the test data is hard coded in the script.

In the data-driven method, the test data is separated from the scripts. This is called parameterization. Parameterized test scripts enable the test data to be extended or changed without having to adapt the test scripts. This enhances the maintainability of the test scripts and the coverage of the automated tests because it's easy to repeat the test with different test data [Fewster, 2006]. The test data is entered in a separate test data repository. Using a test data repository is described in Chap. 14, Test data.

A test data repository can also be used when designing the physical test cases, which is why it is worth defining the test data in such a way that it can be used by the test tool.

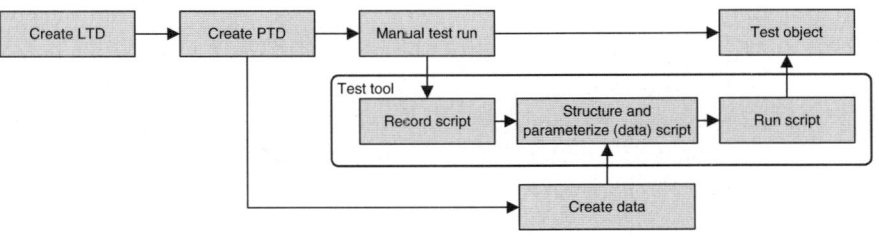

Fig. 16.7 Structured and data-driven scripting

Action-Word-Driven Scripting

In this method, the test tool is driven by action words. Action words are generic procedures that are stored in a library and can be called by the test script. Procedures consist of a group of statements. Executing a statement executes a test action. For example, the action word "log in" that is used when a Web application is tested. The log in procedure ensures that the Web browser is started, that the log in page is opened and that the appropriate user name can be entered. Action words are "building blocks" that enable test scripts to be quickly created and modified.

Physical test cases are separate from the test tool. However, the action words and test data that are used in the automated tests can also be used in the physical test cases. Write the test cases in such a way that the action words and test data are easy to extract and reuse in the test script (see Sect. 13.4 Test Actions).

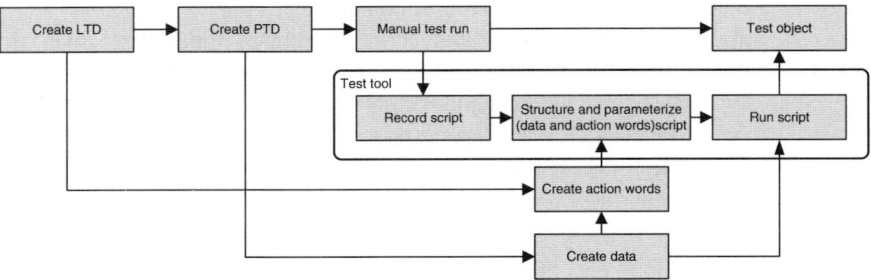

Fig. 16.8 Action-word-driven scripting

16.7.2   Programming Test Scripts

Creating a script using a record and playback tool assumes the availability of a user interface. There are, however, enough situations in which the user interface is not yet available but the automated tests are being prepared. This is the case for smart cards, transaction processing systems and switchboards.

Example 16.1

How should a bank card with a chip, a so-called smart card, be tested? The card is a piece of plastic with a chip consisting of metal contact surfaces. A smart card does not have a user interface; the only way to approach the payment application is through the metal contact surfaces.

Scripts need to be created differently for situations that do not have a user interface. Programming the scripts (scripting) is the solution. The same rules apply to scripting as to developing an application. The test project now has a development project.

Just as record and playback, scripting has different levels of maturity. Test data can be hard coded in a script. To increase the reusability and the maintainability of the tests, a test script library can be used to separate the test data, configuration parameters and test procedures from the test script.

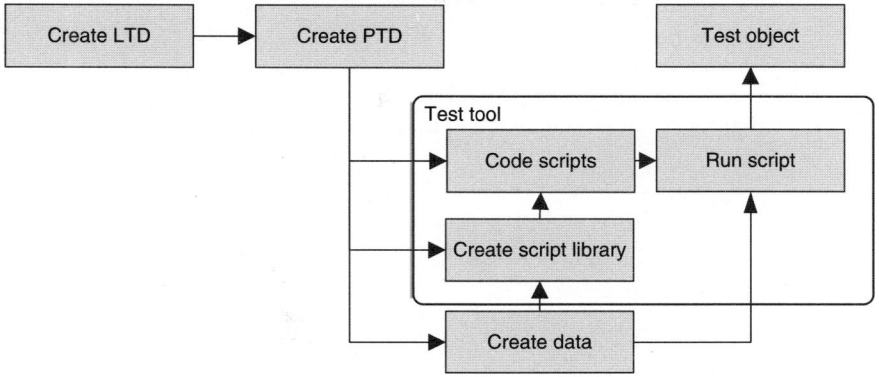

Fig. 16.9 Using a test script library

A well maintained test script library makes it easy to develop a lot of different test scripts. The test script library needs to be well structured. Prior to developing the test script library, it should be decided how the test scripts will be set up and which tests and generic building blocks are needed. This knowledge is then used to define the architecture of the test script library, bearing completeness, accuracy, maintainability and configurability in mind. There are a lot of advantages to a test tool that is easy to configure and control because it enables existing test scripts to be used for different systems. This benefits the reusability of the test scripts and the efficiency of the test process.

> Programmed scripts can build on the physical test design, but it takes a lot of time to first elaborate the physical tests and then script them. If a manual test does not have to be run to create the scripts, the scripts can be based on the logical test design. The script then represents the physical test case.

16.8 Automated Testing for Systems with More Than One Interface

A system can have one or more interfaces. The above-mentioned bank card with the chip is a typical example of a system with one interface.

Most systems do, however, have more than one interface. Systems often have a user interface and a technical interface to the database. Additional interfaces are usually required to input, output and exchange data. An ATM is a good example of a system with several interfaces. The below example shows how a number of test tools can be used simultaneously to run automated tests on the ATM.

Example 16.2: Testing an ATM

An ATM has more than one interface. We distinguish the smart card's interfaces to the user, the bank card and the bank's host systems. Each of the interfaces is used alternately during a transaction. The terminal reads information from the bank card, gives instructions to the user, and checks the balance on the host system.

To test the interface to the card, a test tool is used that intercepts the communication with the card and simulates the smart card if necessary. The tester does not have to rely on physical test cards and can easily create error situations, such as the use of a blocked card. A simulation of the bank system is used for the interface to the host system. The simulator creates test situations such as an insufficient balance or a non-existent account number. The interface to the user is harder to automate. ATMs do not have a Windows interface and they all have a different design; for example, the keypad and the display.

To test the functionality of the ATM, all of the interfaces have to be tested at the same time.

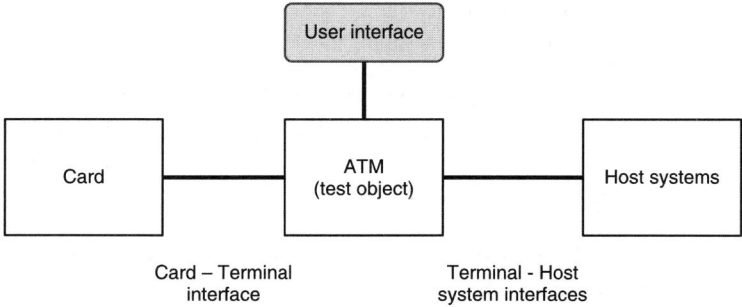

When testing a system with several interfaces, all of the interfaces have to be tested at the same time. A number of test tools are connected to the test object's interfaces. In order to run through a test scenario, the test tools have to be synchronized. The tester can do this himself or automate the task by implementing a controller.

A controller sends instructions and triggers to the other test tools. The controller ensures that every test tool knows which test it has to run, which test data is needed and what the expected system reaction is. The controller can also provide the tester with scripted instructions for the test actions that need to be performed. Building on the above-mentioned example, the tester can receive instructions such as "Press the balance information key" or "Check whether the terminal displays the 'Insufficient balance' message."

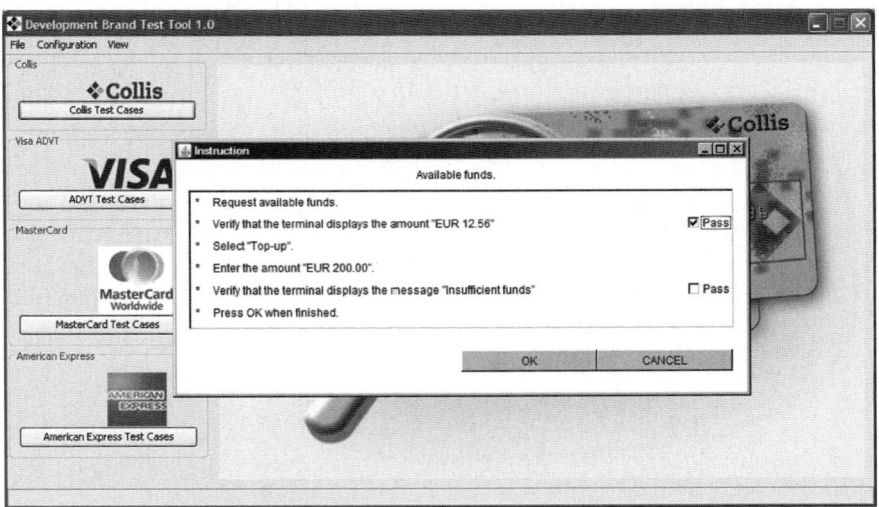

Fig. 16.10 The Conclusion Test Platform® uses user instructions to integrate the manual test actions and the user interface with the automated test run when testing an ATM.

The controller also generates the test report. In principle, each of the test tools can independently generate output. A disadvantage of this is that there will be more than one test report for each test. The test tool architecture can be set up in such a way that a script combines the results of all of the interfaces into one test result. The controller collects the output of the separate test tools and integrates them into one test report.

A test automation architecture is needed to link several test tools as mentioned above. Together, the test tools create an automated test environment that can test all of the interfaces at the same time. The test environment isolates the test object and is therefore sometimes called a "test harness"

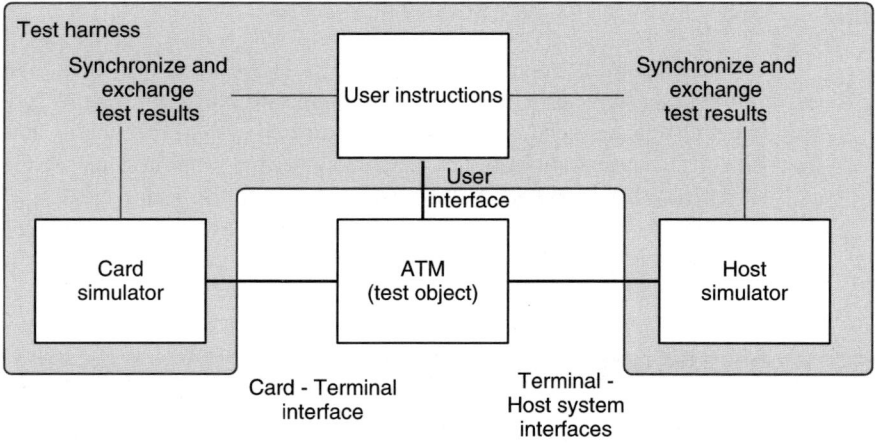

Fig. 16.11 A test harness

A test harness can consist of tools from various suppliers, which is why it can take a while to overcome all of the interface and synchronization problems. Integrated solutions are available to test special applications such as ATMs. The supplier will have already done the integration; this saves a lot of time and benefits the stability of the test environment.

Step 4 – Set up

17   Smoke Test

17.1   Introduction

The smoke test has a lot of similarities with the sanity check; the process of both tests is more or less the same. The main difference between the tests is the time at which they are run and the object they focus on. The sanity check is run earlier on in the test project and focuses on the system design and the testware. The smoke test focuses on the system that is to be tested and is run on the installed and configured test environment (see Sect. 15.5).

The purpose of the smoke test is to determine whether the system is of sufficient quality before it is released for testing. The smoke test is often run more than once because the system is often presented for testing more than once. Based on the experiences of previous deliveries, the smoke test can be more or less rigorous.

The smoke test provides insight into the quality and the testability of the system. The smoke test uses a configured system to determine whether or not the project's products are present. Examples of project products are release notes, the release advice of the previous test phase, and installation instructions. The test also determines whether the delivery is of sufficient quality to start testing: "Does the system start up?", "Does the most important feasible path work?", "Is the new functionality accessible?"

The smoke test is not intended to bring the project to a grinding halt but to make risks discussable. Accepting the test assignment means accepting responsibility. The test budget and planning create an expectation of the number of hours needed to test the system. If the quality of the sys-

D.-J. de Grood, *TestGoal*,
DOI: 10.1007/978-3-540-78828-7_17, © Collis B.V., Leiden, The Netherlands, 2008

tem is lower than expected, the efficiency of the test process will suffer and the agreed deadline may even be at risk. The latter should be communicated on time.

The smoke test provides insight into the degree to which the system meets the expectations and enables the tester to issue a warning if the quality is lower than expected. The measures that need to be taken are agreed in consultation with the customer.

Important reasons to perform a smoke test are saving testing time and maintaining the stability of the test environment. Experience shows that delivered systems can work so poorly that it is not efficient to start the test. If this is not discovered until after the test has started, precious testing time will be lost.

Example 17.1

The development team of the Connecta project (see Example 1.3) has delivered the system. Because the delivery was a little later than expected it's decided not to run a smoke test. David Bloom's test team has prepared the test run really very well and is ready to start. As soon as the system is ready, the testers log in and start running the tests they were assigned.

The first error is quick to surface. While David is talking to one tester about the error, he picks up the following conversation: "Did you know that the system crashes when you import a batch? "You have that problem, too? I can't run my tests without batches!" The fourth tester says he's already logged the error.

It was a bad start. Testing is suspended because David thinks it's a waste of time if all of the testers work on the same error. "I don't care if the system is delivered on time. The next time I'll ask one of the testers to run through the application first. Only if it appears to be working properly will I ask the other team members to start."

The instrument of the smoke test is the smoke test checklist. The checklist is used to determine whether or not the test can start. The products of the smoke test are:

- Errors
  The first errors are found during the smoke test and are logged according to the error reporting procedure.

- A completed checklist containing:
  - A conclusion on the testability of the system.
  - An overview of the measures that need to be taken before the test run can start without risks. If the system does not meet the requirements, the points at which it fails are specified. The factors preventing an efficient test run and the measures that need to be taken to cover them are specified for each point.

The smoke test is a risk-based activity. The TRA describes the components the stakeholders have defined as more or less important. The smoke test focuses on components that need to work in order to run the most important tests.

17.2   Filling out the Checklist

There are two aspects to filling out the checklist. One aspect consists of checking the presence of products, the other of checking the system's workings. The workings are checked by performing a few basic system actions such as logging in, approaching the new function, and writing a record to the database. These checks can be extended by running through the most important test scenario.

In principle, the test engineer who configured the system will execute the smoke test (see also Sect. 15.5). He knows the system, he knows the configuration and he knows the tests that will be run. The test engineer is also familiar with the bug tracking system. From experience, however, we know that the first errors are not caused by bugs in the software, but by errors in the configuration. To check the configuration and any changes made to it, the test engineer runs through the system. By doing this, he's actually doing a smoke test. It's all right that the configuration and smoke test overlap as long as the purpose of the smoke test is not forgotten.

When the system is released for testing, the test engineer runs through the system according to the smoke test checklist. If the system is not good enough, he indicates the points at which the system is failing. For each point, he indicates the risks the failures pose for the anticipated goal and how they prevent the tests being run efficiently. He also suggests measures to cover the risks. Table 17.1 provides a few examples.

Step 4 – Set up

Table 17.1 Measures to cover risks

Error	Risk	Measure
The most important "feasible path" cannot be completed.	The system does not suffice at all, a new release is necessary. Not all tests can be run.	A new release is needed before testing can be continued. The project needs to plan for more releases and a longer testing time.
The system is unstable. The system has a lot of errors.	The tests will take longer to run than expected.	
There are no release notes.	It is not clear which errors have been solved and which known issues are relevant. Functions are tested although they are known not to work, or known errors are being logged. It is not known which changes have to be made to the configuration.	Release notes are still needed. Organize a knowledge transfer between the development team and the test team. The project needs to plan for unnecessary testing time and additional meetings.

The smoke test concludes with the conclusion on the degree to which the system is suitable for testing. The smoke test report consists of the completed checklist, which may contain additional information to support the conclusion. Section 11.2 contains an example of a completed checklist. Although the example represents the sanity check report, it still gives a good impression of the checklist used in the smoke test.

17.3 Maintaining the Checklist

Experiences from the test run, previous smoke tests or other comparable projects can be included in the checklist. Including experiences ensures the smoke test will be run according to the latest insights and that the acquired knowledge is optimally reused.

The experiences from previous test runs can, for example, be used to adjust the most important "feasible path" that is run through during the smoke test. If so wished, the smoke test can be extended with the retesting of the showstoppers from the previous release. Building on Example 17.1, it can, for example, be a good idea to check the importing of batches in the new release. It doesn't make much sense to start testing if the function still doesn't work.

The checklists should be changed according to a controlled change procedure to make sure everyone's using the last version and that it doesn't get too long.

Step 5 – Execution

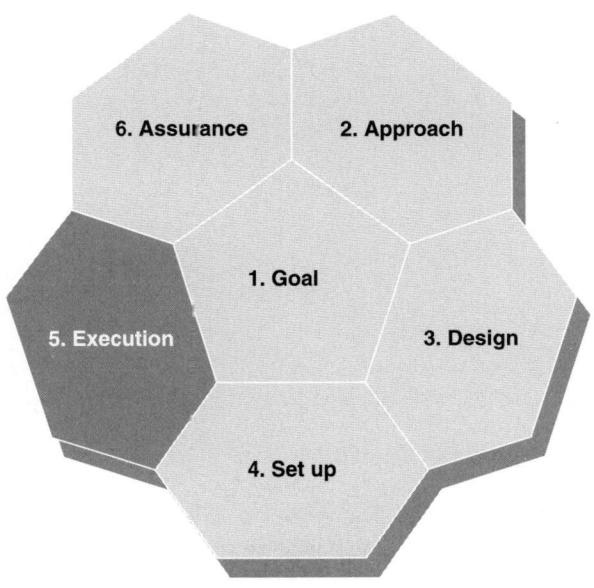

All of the manual and automated tests are run in the Execution step. Errors are logged, solved and retested. The unchanged system components are checked for regression. This step concludes with a release advice that is formulated in the final test report. The Execution step consists of the following activities and products:

Activity	Product
Run tests	Test results
Run retests	Errors
Run regression tests	Test reports
Collect errors	Final test report

18 Test Execution

18.1 Test Execution and its Activities

Remember, the goal of the test project is twofold: to help improve the quality of the system and to make a statement about the degree to which the system will contribute to the anticipated goal. Until the system becomes available, the test activities will focus on understanding the anticipated goal, performing reviews and sanity checks, and preparing a test design and a test environment. The errors found during these activities should help improve the system. In order to make a statement about the real quality, the testers have to work with the system. This they do during the test runs.

Although the actual testing is only a small part of the total test project, it is the most visible of the test project's activities. This is the point in the project at which all the focus is on the testers.

The test execution period distinguishes itself in two ways from the other activities. More than during the preparation, an impression of the quality of the system emerges as the test execution proceeds. The test execution is always very exciting for the stakeholders. When communicating, the tester must realize that others may have a different impression of the activities. Intentionally or not, rating a system also makes a statement about the project leader and the development team. If the test team finds a lot of errors, it will be difficult for the project leader to keep the deadline. The users and business management are eager to know whether the system can go live and are curious about the errors the test team is finding. In many projects, people walk into the test room during testing to try to find out what the chances of a positive release advice are.

D.-J. de Grood, *TestGoal*,
DOI: 10.1007/978-3-540-78828-7_18, © Collis B.V., Leiden, The Netherlands, 2008

Fig. 18.1 Most of an iceberg's mass is below water. In a test project, most of the time is spent on good preparation. But in the end it's the test execution that sticks out [Pol et all,1999].

Be careful with making early statements about the quality of the system and the progress of the tests. The testers are not aware of the context and do not know their audience. A tester will be enthusiastic when a test finally succeeds. The listener will apply such a reaction to the whole system. If this happens, it will be a lot more difficult to tell the stakeholders that the system is not very good. The same applies the other way around. Negative signals about one test that simply won't work can demotivate the user group. "See what I mean? I said it was no good, and now we'll have to work with a bad system," although only one test did not work and the rest of the system is of good quality.

It is important that the test team provides insight into and a good overview of the quality of the test object. This they do according to the procedure and at the moments specified in the test plan (see also Chap. 20 Test Reporting).

The test execution is also different from the preparation activities because it is often in the critical path. The test design is made at the same time the system is being built. A lot of attention is paid to the development process because it determines how the project will progress. This changes as soon as testing starts: the test project now determines the progress of the project. A delay in the test project can have direct consequences on the go-live date.

We have often heard that there is less time for testing when the project is delayed. Fewer tests are run than originally planned, while the deadline remains fixed. This is certainly the case for low-priority tests. Experience shows that if the test team is able to explain the outstanding risks and if the tests have revealed some serious showstoppers, the stakeholders will be willing to wait for the test results. This requires good collaboration between the stakeholders and a good supply of information. Throughout the test execution, provide good insight into the quality of the test object as well as a good overview of the test project's progress.

18.2   Activities During the Test Execution

The test execution is an iterative process, meaning that the activities might be repeated more than once. Every test run includes, but is not limited to, the following activities:

Running the Planned Tests
The testers run the planned tests. The most important tests should be run first in case there's not enough time to run all of the tests. The test coordinator monitors whether the project is progressing well and whether there are bottlenecks.

Logging the Test Results
The result of every test is documented. This produces information about the quality of the test object and the progress of the test project. The ratio between successful and failed tests is a measure for the system's quality. The part of the test set for which the test result has already been documented provides a measure for the progress of the test project. The progress provides insight into when testing will be finished and into the uncovered risks. The more tests that are run, the smaller the chance showstoppers will be found.

Step 5 – Execution

ID	Test action	Expected result	OK	NOK
▶ 1	In main menu select [registration].	Registration wizard is shown to the user.	☑	☐
2	Select [Male client].	The Data-entry window is shown to the user.	☑	☐
3	Enter a valid name (Jâñ). Enter a valid surname (Janşén). Select [Save].	The records are saved in the DB.	☐	☑

Fig. 18.2 The test results are documented

Logging Errors

Every error is logged. Error logging consists of more than just properly documenting an error. If possible, errors should be discussed with the developer [Buchholtz, 2006]. It is not officially the task of the tester to investigate the source of the error, but it is often worth determining whether the error is in the code, the system design or the test case.

However, efficiency is of the essence: it does not make sense for a tester to spend hours looking into something a developer can check in five minutes. Experience shows that it is efficient for a tester to sit with a developer to find the cause of critical errors. This makes estimating the severity and the required solution time a lot more realistic. The fact that the two work together also prevents the developer misinterpreting the error and solving a different problem. The knowledge gained about the source of the problem is of course documented with the error. The test coordinator decides whether an error is submitted for fixing. Chapter 19 describes error logging in detail.

Discussing and Solving Errors

For each of the recorded errors it is decided whether, and how quickly, it will be solved. Ideally, this is done in a controlled manner. Not every error needs to be solved, and not every error is urgent. Error management ensures the development capacity is used for the most important errors. The test plan describes how error management is organized. See Sect. 19.4 for information on error management.

Planning New Versions

If errors in the system make further testing difficult or impossible, the test coordinator will organize a bug-fix version for his testers. This he does by maintaining contact with the project leader responsible for the build or with the release manager. New releases cost time and money, which is why it is advisable to finish the test project with as few new versions as possible. But, an error that makes further testing impossible does have to be solved quickly. This is certainly the case if testing is on the critical path.

Step 5 – Execution

When planning a new version, the consideration is often: "If we get a new version today in which the showstoppers have been solved, we can run the planned tests and retests. This will enable us to minimize the risks. If we carry on testing for a while, we might find more showstoppers that can be directly implemented in the new version." This may slow down the progress of the current test, but it can also save an additional release, and that means saving time.

Reporting Progress and Quality

Reporting plays an important role in result-driven testing. The reports ensure the stakeholders have insight into the test process. The focus on the test project is particularly high during the test run and there is often a need for up-to-date information about progress and quality. A centrally available dashboard containing the fulfills the requirement for up-to-date information.

Error logging and test reports reporting are discussed more extensively in the next chapters.

18.3   Test Run and Regression Tests

During testing terms such as test run, bug-fix versions, retests and regression tests are used. The below figure defines these terms and puts them in context.

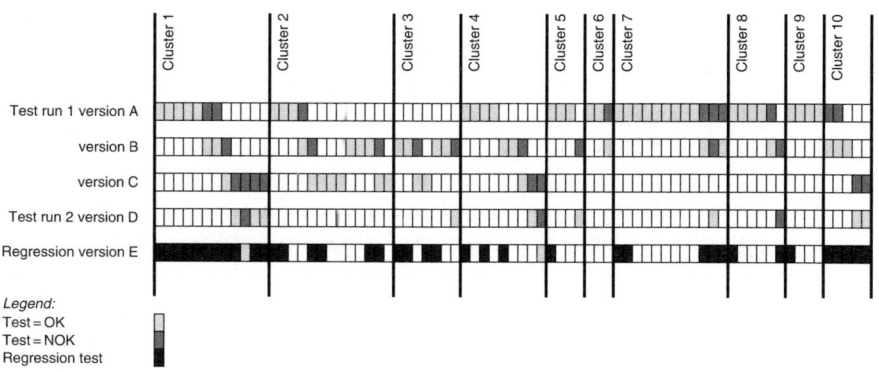

Fig. 18.3 Test execution

Figure 18.3 shows a number of test clusters , the groups of tests that were defined during the TRA and that are also found in the structure of the physical test design. The below describes the work of a tester, so to speak, by looking over his shoulder. Testing has just begun.

When testing started, the tester started the first test run on version A of the test object. The tester started with a cluster in the most important risk category, cluster 1, which consists of twelve test cases. The first tests were successful (OK). The sixth test failed (NOK), but the error wasn't a showstopper so testing continued normally. The seventh test failed too, but this error was a showstopper so the remaining tests in the cluster could not be run. The tester started running the tests in cluster 2. The fourth test in this cluster also produced a showstopper. The tester indicated that he needed a new version in order to finish testing the clusters. While the errors were being processed and the development team was building a new version, the tester tested the other clusters, which are in a lower risk category.

The requested bug-fix version is available; this is version B. The tester checks whether the errors have been solved and tries to finish his tests for cluster 1 because it's one of the most important clusters. The errors have been fixed, but he encounters another showstopper (test case 8). The tester stops testing cluster 1 and continues with the other clusters. He concludes that some errors have been solved and that other errors will apparently be solved in the next version (test case 11 in cluster 7 is, for example, not OK in version B either). He also finds a number of new errors, some of which are showstoppers. So a new version is needed; version C.

All of the reported showstoppers are solved in version C. This does not mean that all of the tests will succeed, but that the tester can run all of his tests. When all of the tests have been run once, the first test run is finished.

> A test run is a compilation of test activities in which all of the tests have been run at least once. However, it is possible that more versions of the system are needed to complete a test run. At the end of a test run, all of the tests will have been run once, but that doesn't mean that all of the tests will have been successful.

All of the tests have been run once, but a new version is needed to check whether all of the errors have been solved correctly. The tester does this using the fourth release, version D. Almost all of the errors have been solved; it is decided to solve two of the three errors in the final version. Version E is used to retest these two errors.

A regression test is also run. Some tests have not been run since version A. The tester would like to make sure the clusters work from start to finish in version E. Because there is not a lot of time, the tester suggests not rerunning all of the tests. Using his experience with the system, he puts together a good regression test set. He selects representative tests for each cluster. He will select more tests for the clusters that contained a lot of errors, and fewer tests for the clusters that did not produce any errors (for example, cluster 9). This will save a lot of testing time.

> A regression test is a test that checks whether regression has occurred. Regression is the phenomenon that unchanged functionality no longer works in a next version of the test object.
>
> A retest is a test that checks whether errors have been properly solved.

18.4   Leaving the Beaten Track

The previous section assumes that only planned tests are run. In some test projects this is an absolute necessity, for example, for conformity tests, during which all of the points described in the standards have to be tested. Thinking up new test cases is not a point of attention. The opposite applies to the testing of new software, where it is difficult to determine beforehand what the solution will look like. Because more experience is gained about the system during each test run, it makes sense to include this experience in the next test run.

In practice we see that new tests are gradually added to the test set or that existing tests lose importance. This means that the test design has to be updated. This is best done between two releases.

Another way of using the experience and knowledge gained about the system is to use exploratory tests alongside the more traditional test design techniques . The execution of the test charter has already been planned, but the actual test is not defined until it's scheduled to run. This enables the testers to use the experiences with the system while designing effective tests and to make the test project more flexible (see also Sect. 12.6.13 Exploratory Testing).

A third option is to reserve time during the test run, for example, on Friday afternoon, when testers can test off the beaten track. The test coordinator encourages his testers to use their system and test knowledge to find hidden errors. The testers are stimulated to leave the beaten track. Depending on the organizational structure, it may be worth introducing an element of fun, for example, by awarding a prize to the tester who logs the most new errors or the first tester who finds a showstopper. Experience shows that this is not only effective, but that it is also experienced as a welcome break from daily routine.

18.5   When is Testing Finished?

The question that is closely connected to running tests is the question of when testing will be finished. The next question that is asked in almost the same breath is who determines when the test project is finished.

In principle business will decide whether the testing activities can be stopped or should be continued. The role of the test coordinator is to provide business with information so it can make a well-informed and correct decision (see Chap. 20 Test Reporting). The following, correlated issues should be considered.

All of the Planned Tests Have Been Run
If all of the defined tests have been run, it can be established that the test work has been done. That is to say, if all of the tests have finished successfully and the test team and stakeholders feel that everything has been tested with the defined test set. The chapter on the test plan describes how a test design review can be performed. We have seen that this review helps the stakeholders assess the tests and the release advise (see Sect. 10.4.8). Involving the stakeholders in the test project at an early stage during the review adds value. For automated test runs, it can occur that all of the tests will have been run. For manual test runs, this is hardly ever is the case. There is usually one more test a tester will want to run. But let's be honest, finished is finished. The information needed to determine whether all of the tests have been run can be found in the test report (see Sect. 20.2.10 Test result by test risk category or test cluster).

Unsolved Errors
Are there errors that have not been solved or that need to be retested? If this is the case, testing is not finished. Unless, of course, it has been decided that the unsolved errors are not showstoppers, in which case test-

ing is finished. The information needed to answer this question can be found in the test report (see Sect. 20.2.7 Error status).

A New Showstopper
Even if all of the known errors have been solved, it does not mean that the system is free of errors. Further testing can produce a new show-stopper. The defect detection rate (see Sect. 20.2.8) provides insight into the time that is expected to be needed to find the next error. This time can, converted into costs, be weighed against the expected damage if a showstopper is found in the live system. If the defect detection rate is low, it may be more lucrative to stop testing.

Outstanding Risks
As shown in Example 7.3, testing minimizes risks. The further the test execution progresses, the more outstanding risks will be minimized. At set times, business will have to consider the following: What is the expected damage if the outstanding risks occur, and what will it cost to minimize these risks with testing? It is common to stop testing before all of the risks have been minimized. The outstanding product risks can be found in the test report (see Sect. 20.2.12 Outstanding Product Risks).

In addition to the above-mentioned "objective" consideration, there are always some considerations that are less tangible but still influence the decision, such as those mentioned below.

The General Feeling About the System
During testing, the testers and the stakeholders develop an impression of the system's performance. This feeling is often key in the decision to stop testing. If all of the tests have been run but the general feeling is not good, testing will often continue. Conversely, it is also true that if the general feeling is good, the stakeholders will be inclined to attach less importance to outstanding issues. This is all right, as long as the testers take their responsibility by feeding objective information into the decision-making process.

The Importance of Going Live
Regardless of facts or feelings, going live can be so important that it negates all arguments. The system goes live with or without errors, with or without risks. The tester has to be able to recognize this situation and adapt the test strategy. In this situation, the added value of testing is not the well-considered release advice, but minimizing the number of errors in the live environment.

It is result driven not to give a release advice when this does not
has any added value.

 Note that arguments such as "no money" and "no time" have not been
taken into consideration. Even though these arguments are often men-
tioned, experience shows that if the risks are too high additional time
and money will be provided. Whether this really happens depends on
the outstanding risks and the confidence business has in testing.

19   Error Logging and Management

19.1   Introduction

Errors are found during the test run. Logging the errors in a disciplined way provides insight and makes it possible to solve the errors efficiently. A prerequisite is that an error should be unambiguous and legible and contain all of the information needed to process and manage it. This chapter discusses how errors are logged to ensure the requirements are met. The information in this chapter can also serve as a guideline for the organization and as a starting point for the testers. Clearly describing how errors are logged saves a lot of time when discussing and solving them.

An error is found as soon as something is identified in the test base or in the test object that deviates from the expectation. Errors are found in:

- The test base
- The programs
- The test design
- The test data
- The technical infrastructure

Findings constitute an observed difference between expectation and outcome; they indicate where the anticipated goal is at risk. A bug in the code is the most well-known occurrence of an finding, but not every finding is necessarily a bug. An example of this is the enhancement, often called change or RfC (Request for Change); the system is working according to specification, but the tester has a suggestion for improvement. Another example is the test error; the tester logs an defect during a test and closer analysis reveals that the defect is not in the test object,

D.-J. de Grood, *TestGoal*,
DOI: 10.1007/978-3-540-78828-7_19, © Collis B.V., Leiden, The Netherlands, 2008

Step 5 – Execution

but in the designed tests. That is why the he tester must be diplomatic when reporting errors and take into account that he too can be the cause of an error. Besides that, emphasizing that the supplier or the developer has made "yet another mistake" does not benefit communication.

The importance and purpose of good error logging are:

- Errors can be checked, traced and reproduced

- Priorities can be defined

- Insight into the quality of the test object is provided

Simply logging errors does not have much added value. A result is not achieved until the errors have been processed. How this is done, depends on how error management has been set up. To ensure error management runs smoothly, it is important to log the errors properly, accurately and completely.

19.2   Filling out the Error Log

Errors should always be logged to provide insight into the quality of the test object. Considering that the aim is good quality software, good logging and the controlled processing of errors (including bug fixing) is essential. Accordingly, error logs must meet a certain standard. An error report is used to:

- Indicate the risks for the anticipated goal if the test object goes live.
- Process errors by implementing the solution in the documentation or the code. This requires clear references and an unambiguous, complete description.
- Retest the error. It must be clear which test or test actions caused the error.

Time pressure can make it tempting to "quickly" log the error, but it is in the interest of the whole organization to log errors accurately and completely.

Experience shows that for every five minutes the tester saves on error logging, a multiple of this time is lost at a later stage. This is due to discussions and explanations that need to be given, incorrectly solved errors and the retesting of incorrectly solved errors. In many cases, the tester will have to show the developer how he produced the error. Always take the time to log an error and include this time in the planning.

A lot of organizations have resources or tools that can be used to generate and view error logs. Sometimes reports and statistics are automatically structured using the attributes that were added with the error. To ensure the reports are actually trustworthy, testers have to be disciplined when it comes to entering information. It is not useful (and can even have the opposite effect) to document certain attributes if they are not entered properly.

Non-reproducible errors should also be logged because they can turn out to be important at a later stage. Do not wait too long before logging an error or you'll forget it. Additional observations or information can always be added at a later stage.

19.3 Error Attributes

The following attributes are logged together with an error. The last column indicates whether the field is a "must have" (A) or an optional enhancement (B).

Table 19.1 Error attributes

Attribute	Description	A/B
Number	Unique identification number of the error.	A
Date	Date on which the error was found.	A
Test level and test object	Test level at which the error was found, for example: • System test of test object A • User acceptance test of test object B If the test level is obvious, this field can be omitted. This can be the case if project only has one test level.	B

Table 19.1 Continued

Attribute	Description	A/B
Type	Type of error, for example, an error that occurs in: • The test base • The code • The test design • The test data • The technical infrastructure Keeping track of the error category enables targeted improvements to be made to development processes. If the organization does not analyze errors, this field can be omitted.	B
Tester	The tester who logged the error.	A
Error status	The status of the error indicates whether an error has been found, solved or closed. The actual status descriptions that are used will depend on the standard that is applied by the organization. Possible statuses are: **Found** The error has been logged and needs to be solved by the development team. **Solved** The error has been solved by the development team and needs to be retested by the test team. **Closed** The test team has retested the error and the problem does not occur anymore, or the problem was incorrectly logged and does not need further attention. If needed, additional statuses can be added, such as: **Duplicate Record** The error was logged twice, this error is not processed. **Rejected** The error was unjustified. The test object is working as it should or the error is a result of an incorrect assumption made by the tester. The error is not processed. **Assigned** The error has been assigned to a developer/analyst but has not been solved yet.	A

Table 19.1 Continued

Attribute	Description	A/B
	Reopened The error was solved and retested by the test team. Unfortunately, the solution was incorrect or incomplete, so the error has been reopened. If the organization has little experience with error logging, it is advisable to limit the number of statuses.	
Severity	The severity of the error. There are three categories: High, Medium and Low. **High** The test object cannot be tested. The functionality of the test object or of the main component is not working properly. This means that other parts cannot be tested and that there will be no test results. Errors in this category are also called showstoppers. The error is so severe that the application cannot go live. Examples of showstoppers are primary processes that cannot be executed, a very unstable system, or errors in key calculations. **Medium** The test object cannot be tested. One or more functionalities are not working as they should but are not stopping further testing. The error has no consequences for the other test results and the progress of the project. This category is also called serious. **Low** All of the functions that do not influence the functionality of the test object. Examples of this are typos and GUI errors. This category is also called cosmetic. If needed, more categories can be added, for example, Nice to have. In practice, a classification in three categories works well. Limiting the number of severity levels saves time when assigning and discussing errors. Too many levels muddle the distinction between the severities.	A
Urgency	In addition to the severity of the error, the urgency of the error also determines when it will be solved. The field "urgency" is most important when there are a lot of errors with a certain severity.	B

Table 19.1 Continued

Attribute	Description	A/B
	Urgency then determines the sequence in which the errors are processed. Urgency is important for the software development project and does not make any statements about going live. *Example:* *In a project, all of the errors with the code "high" were solved. There are 120 errors with the code "medium," but there is not enough time to solve all of them. The urgency gives the developer insight into the errors he needs to solve first.* As shown in the example, this field is especially important if the test project is nearing its end and errors need to be prioritized.	
Test design	A reference to the test case to which the problem notification relates. In practice, a lot of errors are not directly connected to a test case. They are found during the execution of a test case but they identify another error than the one the test case was designed for. This attribute can sometimes be left empty.	A
Configuration/ release version	Identifies the configuration of the test environment and the version or build number of the test object. This field makes it possible to research version dependency. If external suppliers are used, the name or the code of the supplier must be included in the version description.	A
Fix version	Identifies the configuration of the test environment and the version or build number of the test object in which the reported error is solved. An error that is solved in a version that has already been released can be included in the retest. The field can also indicate a future version in which the error will be solved. After priorities have been set, it is possible that an error will not be solved in the next version. This can be the case if a quick release has to be made because a showstopper is preventing further testing. Quick releases are commonly limited to solutions for showstoppers.	A
Test environment	Identifies the test environment in which the error has been identified. This makes it possible to research environment dependency. This attribute can be omitted if only one environment is used in the project or organization.	B

Table 19.1 Continued

Attribute	Description	A/B
Assignment	The employee or division to which the next action that is linked to the error has been assigned. This can be the analyst who has to research the error, the programmer who has to solve the error or the tester who needs to retest the error.	B
Summary	The title or summary of the error provides a short description of the nature of the error. In the summary, indicate: • Where the error occurs • What the wrong result is • Which test action produces the error *For example:* *NAR screen. Entering a name with diacritical characters produces the error: "Invalid name."*	A
Description	The description should contain all of the information needed to reproduce the error. The description consists of • The test actions executed by the tester • The observed system response • The expected system response • The reason why the system response is incorrect • The impact of the deviation *Example:* *Test action:* *Select [male customer] in the Registration wizard. The NAR screen is displayed. In the NAR screen, enter first name: Jâñ, surname: Janşén Select [Save].* *System response:* *The error message "Invalid name" is displayed.* *Expected response:* *The expected response was that the name Jâñ Janşén would be saved in the system without producing an error message.* *Why this is not correct:* *The characters that are used are not allowed according to the character set specified in FO_CHARSET_V01.*	A

Table 19.1 Continued

Attribute	Description	A/B
	Impact:	
	Certain customer names cannot be entered in the system. The current customer record contains a lot of names with special characters.	
	A reference to the used specifications (including version number) can also be added.	
Comment	Notes from those involved on what was done with the error; additions or actions that need to be performed. Most tools automatically add a name and a date/time to the comment. If this is not the case, start every new comment with, for example, the date followed by the name of the tester: "YY-MM-DD <NAME>:" This field can also be used to indicate the possible solution and to keep track of the error's history.	A
Attachments	If it helps clarify the error, add an attachment with (examples of) system logs, displayed messages or screenshots. A reference can be added instead of an attachment. If there are no attachments, this field can remain empty.	B

Starting the summary with the location where the error occurs, the NAR screen in the above example, enables errors to be quickly located in the report. Sorting the report by description automatically groups the errors.

19.4   Error Management

As previously mentioned, error logging is not a goal in itself; something has to be done with them. Error management ensures that the right errors are processed and closed in a controlled manner. Efficient error management requires good procedures and an error logging process that supports them.

Error management can be set up as described below. The below set up is based on a centralized bug tracking system. The tool enables errors to be assigned to project employees and automatically sends them a notification.

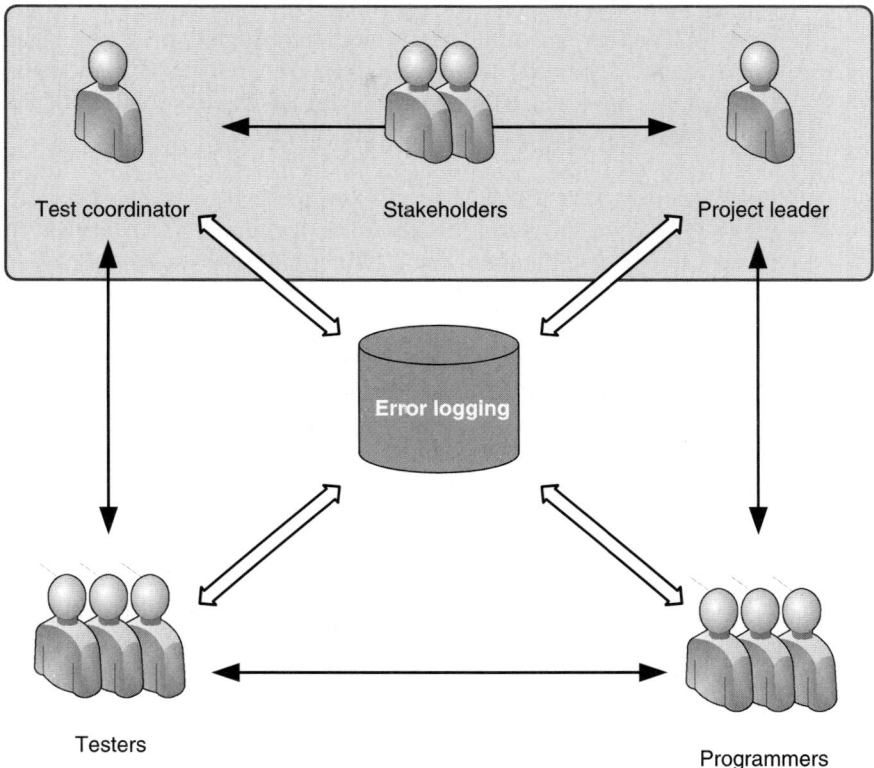

Fig. 19.1 The error management process. The black arrows indicate the information exchange between the people involved. The white arrows indicate that all of the people involved request information and add it to the central error log.

Error management includes the following steps:

1. Logging
A tester observes a system response that is different from what was expected. He examines the situation and repeats his actions in order to check whether the system response can be reproduced. Depending on the type of error, it may be worth the tester discussing it with the programmer [Buchhultz, 2006]. Experience shows that such discussions improve the quality of the error log. The tester logs his error in the bug tracking system, gives it the status "Found" and assigns it to the coordinator.

2. Control

The test coordinator checks that the error has been logged according to the agreed standards. He makes sure the description is clear and checks that any necessary attachments have been added. If this is not the case, he will ask the tester to correct the entry. The test coordinator also checks whether a similar error has already been logged. If this is the case, he indicates this in the error entry, for example, by giving it the status "duplicate record." In this case, he makes sure that any additional information is also added to the earlier entry. If the error is clear and the entry complete, he passes it on to the project leader.

In an organization in which the testers are used to logging errors according to the agreed standards, this control is less important. However, to prevent sloppiness, it is advisable to control the entries at set intervals.

3. Triage Meeting

It does not go without saying that every error is processed by the programmers. If the project is under a lot of time pressure, it is important that the available time is used to solve the most important errors. This is why triage meetings (or error meetings), in which the test coordinator and the project leader discuss the errors, are held at set intervals. The following is usually discussed during these meetings:

New Errors

The desired follow-up action and urgency are determined for each new error.
The test or business impact is determined and it is decided whether the error can be solved with or without adapting the test base or if further analysis is required.

Open Errors

The progress of the unsolved errors is discussed. If the source of the error has been analyzed and it turns out that more than one solution is possible, the solutions can be discussed.

Solved Errors

The retest date is specified for each of the solved errors. Note that, depending the deployment cycle, a solution might not be instantly available on the test environment.

The test coordinator and the project leader do not decide which errors are solved: this is the privilege of the stakeholders (the customer, the acceptor, the user representative, the system administrator), who also participate in the error meetings. However, this does not mean that every error meeting has to be attended by everyone.

At the beginning of a test project, the emphasis is on the impact an error has on testing. The test impact indicates the degree to which testing is hindered by errors in the system. The impact on the test is determined primarily by the test coordinator, not the stakeholders. In the beginning, the test coordinator and the project leader manage the errors together. At the start of the test project, the errors with the biggest impact on the test will be processed first. If the test team cannot run the tests, the most important business scenarios will not be tested. As testing progresses, the emphasis will shift to the business impact. At the end of the test project, the stakeholders will play an important role and will indicate which errors have to be solved before they accept the system [Pinkster et al, 2004].

4. Implementing the Solution
The project leader assigns the error to an analyst or programmer who solves the error. If necessary, the analyst or programmer meets with the tester who logged the error. In some organizations, the developers are isolated from the testers. This is a pity because collaboration between testers and developers helps quickly and effectively solve errors. If the tester demonstrates how he found the error, the programmer can test the solution himself. It may even be possible for the tester to watch the programmer test the solution. This way of working may take more time but the payback is manifold because the retest runs smoothly and the error is solved correctly the first time around.

When the programmer checks in his solution, he changes the error status to "solved" and assigns the error to the project leader who adds the solution to the planned release and informs the test coordinator that the solution is available for retesting.

5. Retesting
When the changed software, which contains the solved errors, is available in the test environment, the test coordinator orders a retest. The retest is preferably executed by the tester who logged the errors. The retest enables the tester to decide whether the error can be closed. If the solution is not correct or if it is incomplete, the error will be represented.

20 Test Reporting

20.1 Introduction

TestGoal focuses on the goal. But it's not enough to give the goal a prominent place; it's also important to inform the stakeholders about the degree to which the goal has been achieved. This is done using the test report, which provides an overview of and insight into the progress of the test project and the quality of the test object.

Virtually every test project uses a progress report. It is also common to write an official test report at the end of the test project. The test report provides a comprehensive summary and contains a conclusion. In most cases, the test manager formulates the conclusion carefully, which is why it is not unusual for the final test report to contain new information that places the release advice in a different context. For the receiving party, this is undesirable and can come as a surprise.

Business, project and test management need clear performance indicators that answer the questions that management has to deal with. The testers who create the reports need to know what these questions are and need to make sure the information provided answers them. This applies throughout and after the test project.

In TestGoal, the progress report and the test report are combined. The result is a test report that provides an overview of and insight into the quality of the test object and the progress of the test project.

The test report is a snapshot that compares the achieved goal to the anticipated goal. Based on the test report, the recipient can decide whether action is necessary. If everything is going according to (test) plan, no

D.-J. de Grood, *TestGoal*,
DOI: 10.1007/978-3-540-78828-7_20, © Collis B.V., Leiden, The Netherlands, 2008

Step 5 – Execution

action is necessary. But in many cases action is necessary and priorities have to be shifted or new actions have to be defined. These actions can apply to the project as well as to the product. At the end of the test project, the last version of the test report functions as the final release advice or report. Although the customer determines the information that a test report should contain, it possible to specify the information that is usually relevant for each step. The test report usually contains the following elements:

General – during all steps

- Scope
- Release advice
- Hour estimate
- Project risks/bottlenecks
- Status of the products

Some information is only relevant for a specific step, which is why the report is occasionally extended with the following elements:

Step 3 – during the test design

- Test designs – realized versus planned tests
- Outstanding errors – for example, in the test base

Step 5 – during the test run

Test results by risk category, test cluster , function or quality attribute

- Errors status
- Defect detection rate
- Outstanding errors
- Test progress – run versus planned tests
- Outstanding product risks

The reports contain information that is extracted from the test process and presented in tables, lists and graphics. Graphics often provide quick insight into the progress, but the essence of reporting is to explain the graphics. The following sections show how report elements can be displayed as graphics and how the graphics can be interpreted.

20.2   Elements in the Test Report

Stakeholders don't always know which reports they want to have. The tester then has to choose what he will report on. This chapter helps him make this choice by describing the elements a test report can consist of. The aspects, the function and the information that can be obtained from a report element are described for each of the elements. This not only helps the tester choose, it can also serve as a manual for the stakeholders that enables them to interpret the test report.

> Make sure the test report is as clear as possible, and report only on the project's key performance indicators. The creation of a test report is never a goal in itself.

20.2.1   *Scope*

The scope clearly defines the product the release advice applies to and what the advice is based on.

Test Object
Identifies the system that is to be tested. Specify the modules, versions and build numbers of the systems that were tested.

The Anticipated Goal
The anticipated goal in terms of insight into and an overview of the working and the quality of the software. The anticipated goal also contains the minimization of uncertainties so that the trust in the software is maximized and the "fit for purpose" guaranteed. This information appears in the Goal description that was created at the beginning of the test project (see Chap. 6 Assessing the Anticipated Goal).

Test Base
Identifies the used test base. Include a reference to the specifications the tests are based on or to whatever the test was run against.

Test Design
Identifies the used test design. Include a reference to the physical tests that were run.

Step 5 – Execution

20.2.2   *Release Advice*

The release advice is the most important output of every test project. The test manager has to indicate whether or not he thinks it makes sense to move the test project into the next phase. The next phase can be a next test level, a pilot or the live environment. The release advice will mainly be based on the status of the product compared to the anticipated goal. This is specified according to

- The outstanding product risks
- The tests that have not been yet been run
- The quality gap

> The quality gap is the difference between the anticipated quality and the actual quality. This difference is often measured by the number of outstanding errors. The difference between the errors found and the errors closed indicates the known gap between the anticipated quality and the actual quality.

The release advice can read as follows:

Positive
The test team has sufficiently tested all of the risk areas and the system meets expectations. The advantages of further testing do not outweigh the risk of finding errors in the live environment.

Conditional
The product can be released, but only if the specified conditions are met. A conditional release advice can be given if, for example, not all of the tests have been completed yet but it is not expected that errors will be found that influence the decision. A conditional release advice can also be given if it is decided to take measures that offset the product risks.

> *Example 20.1 Conditional release*
>
> A conditional release advice is given on Thursday. This advice is the green light for the maintenance department to start preparing the deployment. In the meantime, the remaining tests are run, but it is not expected that anymore showstoppers will be found. The

> conditional release advice is changed to a positive release advice when all of the tests have been completed and no showstoppers have been found.
>
> The system test reveals that two showstoppers have not been solved. Even so, the test manager decides to give a conditional release advice for the pilot so the users can start getting used to the new system. A prerequisite is that the errors are listed in the pilot script so that the users know that two functions are not working as they should. It is also decided that the showstoppers have to be solved within four days and that the solutions are implemented in the pilot environment over the weekend. The system test team will test the bug fix, after which the pilot can be continued as planned.

Negative

Considering the quality and the outstanding product risks it is not wise to release the product for the next phase.

The release advice should supplemented with a suggestion for follow-up actions, i. e. actions that are necessary to achieve a positive release advice. It may also be desirable to make suggestions for a positive release advice.

The release advice is usually negative at the beginning of a test project because the tests that cover the identified product risks still have to be run.

20.2.3   *Hour Estimate*

The graphic in Fig. 20.1 provides an overview of the number of hours estimated for the test project. The graphic consists of three elements.

Number of Hours Estimated
The number of hours estimated is the number of hours agreed for the test project. This number is the same as the hour estimate in the test plan.

Number of Hours Spent
The actual number of hours spent can differ from the expectation. Activities can take more or less time than expected or unforeseen activities

Step 5 – Execution

may have to be carried out. The number of hours spent can usually be traced using the project or organization's hour administration.

Estimate to Completion
The estimate to completion (ETC) is an assessment of the number of hours needed to achieve the agreed result. The ETC includes the experience of the previous period. For example, an activity is estimated to take 10 hours, but 8 hours were needed to complete the first half of the task. It is unlikely that the other half can be completed within the 2 hours that are left. It is more likely that the ETC will be 8 hours and that the task will take a total of 16 hours.

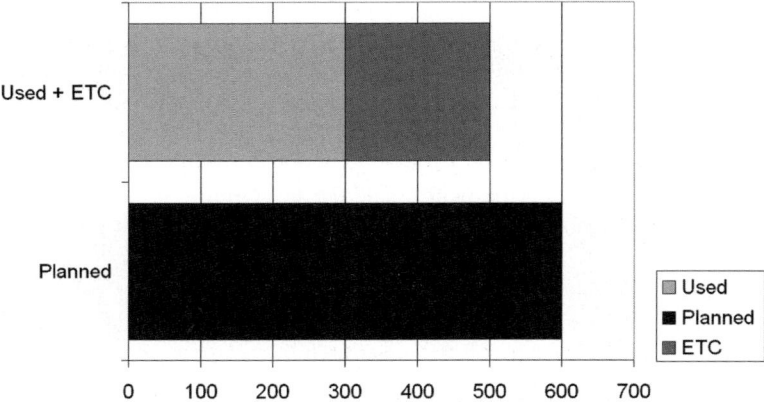

Fig. 20.1 An hour estimate for the test project

The above graphic is a snapshot of the planned time in relation to the time that was spent at that moment and the time expected to be needed. If "Spent + ETC > Estimated," it is realistic to expect that the project will take longer than expected. In this case, the customer should be given a choice of corrective measures.

20.2.4   *Project Risks and Bottlenecks*

The test report contains an overview of the risks and bottlenecks. State the most important risks or bottlenecks that have occurred and jeopardize the success of the project, and discuss them during the progress meeting. Include improvement suggestions when reporting the bottlenecks and risks. You can then decide which measures need to taken during the progress meeting.

20.2.5   Product Status

Indicate the status of the products that are stated in the WBS, for example, using a percentage, a code "Ready Y/N" or the ETC. The ETC is preferable because it provides the manager immediate insight into the time that will be needed. A completed product is indicated with an ETC of 0 hours.

Quality assurance is an important factor in any (test) project. The product overview can specify whether products have been reviewed and accepted. Including these points in the overview ensures that the quality is monitored and prevents documents never being completed.

Product	ETC	Review completed	Accepted
Product A	8 hours		
Product B	0 hours	Y	Y
Product C	0 hours	Y	

Product	Ready	Review completed	Accepted
Product A	N		
Product B	Y	Y	Y
Product C	Y	Y	

Product	Status	Review completed	Accepted
Product A	60%		
Product B	100%	Y	Y
Product C	100%	Y	

20.2.6   Completed Versus Planned Tests

The tests are designed in step 3 of the Planning step. While creating the design, additional reports are generated for the generic points of the completed and planned tests.

Tests are structured in functional clusters that are derived from the test tree that was defined during the TRA. The test cluster consists of, for example, a function, a scenario or a covered risk. The progress of each cluster is indicated in the Design step. Just as for the products, progress can be reported as a percentage, a code or as an ETC (see Sect. 20.2.5).

Step 5 – Execution

When using percentages, make clear agreements on how they are defined; does the percentage say something about the number of components that have been completed or about the amount of work that has been completed? Pareto's famous 80-20 rule [Pareto] says that 80% of the work is done in 20% of the time. Bear in mind that if 80% of the work is reported completed, it doesn't mean that 80% of the budgeted time was used. It's preferable to use ETC because it is the best way to indicate how much work is needed to complete the task. If a report needs to specify percentages, base them on the ETC.

Example 20.1

An example of using ETC as starting point:

Percentage = Hours Spent / (Hours Spent + ETC)

Explanation: In this formula, the number of hours spent is divided by the total number of hours expected to be needed to complete the task. If the task has been estimated correctly, Hours Estimated equals Hours Spent + ETC. The percentage will then be Hours Spent/Hours Estimated.

In other cases, the percentage is adapted to new insights. For example, a task is estimated to take 100 hours. After spending 60 hours on the task, it is estimated that another 50 hours will be needed to complete the task. In this case, the percentage will not be 60/100 = 60%, but 60/(60+50) = 60/110 = 55%

20.2.7   Error Status

In Fig. 20.2, the cumulative number of errors is specified for a date. Errors have the status Found, Solved or Closed.

Found
This curve indicates the total number of errors reported.

Solved
The total number of errors solved. These are errors that have been solved by the developers and will be presented for retesting.
(Make sure the retest is included in the planning)

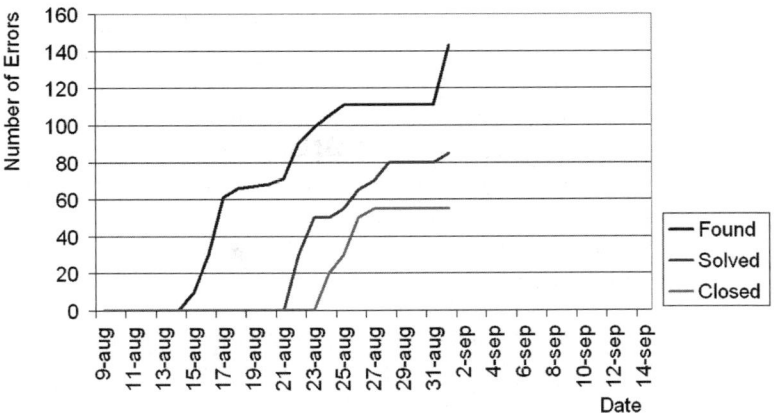

Fig. 20.2 An error status overview

Closed
These are the errors that have been retested and approved by the test organization.

The graphic provides information about the number of errors found, making it an indicator of the quality of the system. The difference between the lines "Found" and "Closed" is indicative of the difference between the anticipated quality and the achieved quality: the "quality gap." In practice, this quality gap will never be completely closed. This means that there are risks for the live environment that have to be explained in progress meetings and in the release advice.

The graphic indicates where the workload lies. The difference between the curves "Found" and "Solved" means that the workload lies with the development team. The difference between the curves "Solved" and "Closed" means that the workload lies with the test team.

The curve *Found errors* indicates how many errors have been logged. Theoretically, this curve is an S-curve: It's the beginning and the test has yet to get going. Because the testers are not very familiar with the system and configuration errors can slow down the test run, there will be a limited number of errors logged per day. As the test gets going, the number of errors will increase and the curve will become steeper until the errors that are easy to find will have all been found. As it will take more and more time to find new errors, the curve will decline.

In practice, the curve will be jagged, because the same amount of testing will not be done every day, the composition of the team may change, or a new release that works slightly differently will produce fewer errors. On the other hand, a new release can also introduce new errors, which can increase the number of errors logged.

The graphic can provide a lot of information about the progress of the test project. The following aspects of the *Found* curve are interesting to monitor: no new errors are reported, or a lot of errors are suddenly reported.

No Errors Are Reported

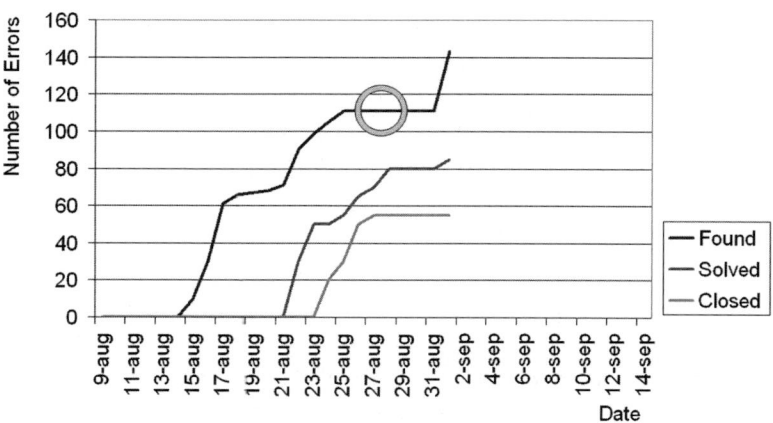

Fig. 20.3 No errors are reported

If no errors are reported, the "Found" line is horizontal: the number of found errors stays the same. The fact that no errors are reported can have a number of reasons:

- The quality of the system is such that it is difficult to find new errors. In this case we say that the "defect detection rate" is low (see Sect. 20.2.8). This can indicate that the test is nearing its end because the cost (= time) of finding the next error will be higher than the cost of fixing it later.
- No errors are found because no testing is done. This can have a number of reasons:
  - No resources are available due to illness or vacation.
  - No resources are available because other activities were assigned a higher priority.
  - No more testing can be done because errors have been logged that make running further tests impossible. The test team is waiting for a new release.
  - No more testing can be done because there are problems with the test environment.
- A lot of testing is done, but the testers aren't very good.

A Lot of Errors Are Suddenly Reported

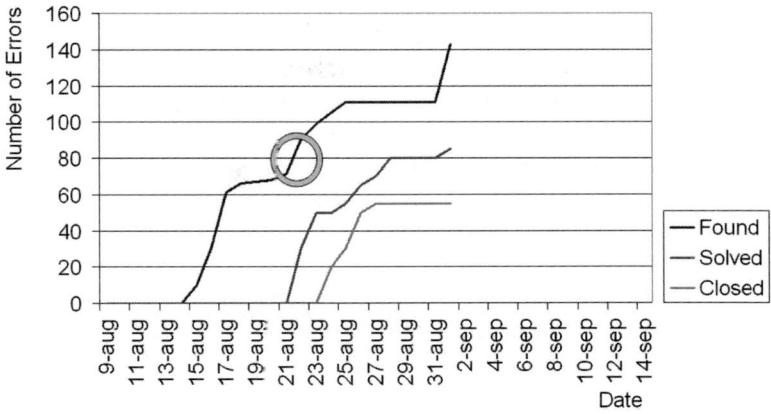

Fig. 20.4 A lot of errors are suddenly reported

A number of factors can cause the "Found errors" line to be steep:

- A new release is delivered that makes it possible to carry on testing. The tests are run on a poorly written function.
- A new release is delivered, but bug fixing introduced a lot of new errors.
- There is an error in the configuration or the deployment of the software, which causes the system to function inadequately.
- More testing is done or more resources are available for the project.
- There was a change in the test team or the team has become better at finding errors.

The "Solved errors" curve indicates how many errors have been solved. This curve should, with some delay, follow the "Open errors" curve. This may not be the case if

- The development department does not give priority to or does not have time to solve the errors. As soon as a release is handed over to the test team, the development team's priority shifts to the new release. The result of this is that error solving is not given enough attention [Clermont, 2006].
- The errors were reported but were not flagged as "to be solved." This suggests that the error management process is behind on the test process.
- A graphic is used that only indicates the errors that have a high priority; only low priority errors are solved.

The curve "Closed errors" indicates how many errors have been closed after retesting. In principle, this curve should be the same as the solved errors curve, but with some delay. This may not be the case if

- The test team does not give priority to or does not have time to retest the solved errors.
- The errors were solved but the bug fix release is not yet available.
- Retesting shows that the errors have not or have only partly been solved. The errors were reopened after retesting.

The graphic can be used to display all of the errors or only the errors in a certain category, for example, a graphic that displays errors with "Severity = High." Include only the most important errors in the management reports.

Curves can be added for other errors statuses, such as "Rejected errors" or "Errors reopened after retesting." The "Rejected errors" curve provides insight into the quality of the logged errors. If a lot of errors are rejected, it may mean that they were wrongfully or inaccurately logged. The number of reopened errors provides insight into the efficiency with which errors are solved.

Be careful when adding additional curves. Make sure the message does not get lost in the huge amount of data that is presented. Keep graphics as simple as possible. Experience shows that the chosen three curves are a good basis for test reporting.

20.2.8 Defect Detection Rate

The defect detection rate indicates the number of errors found for each time unit of testing as a function of the date.

The graphic in Fig. 20.5 indicates how many new errors are expected to be found for a given point in time if testing continues for an hour or a day. This graphic can be used to determine whether further testing is useful. If the cost of finding an error is higher than the expected cost of fixing the error in the live system, further testing is open to debate. Conversely, the project manager will not stop the test if the test coordinator proves that an additional day of testing will most probably produce another eight showstoppers.

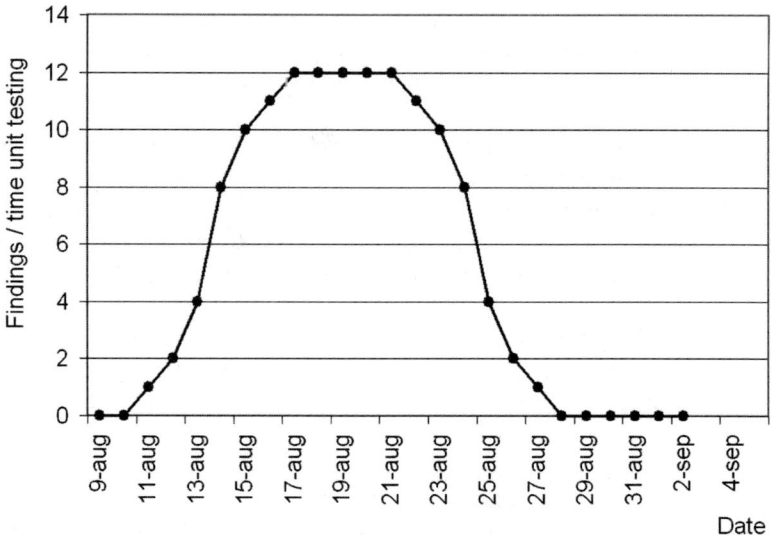

Fig. 20.5 The defect detection rate

The graphic can display all of the errors or only the errors with a specific priority. It can show new errors as well as solved and closed errors.

A prerequisite of using this graphic is that the hour registration is accurate. The hour registration should indicate the real number of hours spent running the tests. All of the testers need to record their hours in the same way. This is not always easy to do, meaning that the graphic is not always accurate. It is, however, a useful guideline.

20.2.9   Open Errors

In addition to an overview of unsolved errors, it's often desirable to gain insight into them.

Logging all of the errors creates long lists. It's best to attach them to the test report rather than include them in the test report so that the test report remains as simple and clear as possible. In this case the test report will contain high-level information, such as an overview of errors by severity or cause.

Step 5 – Execution

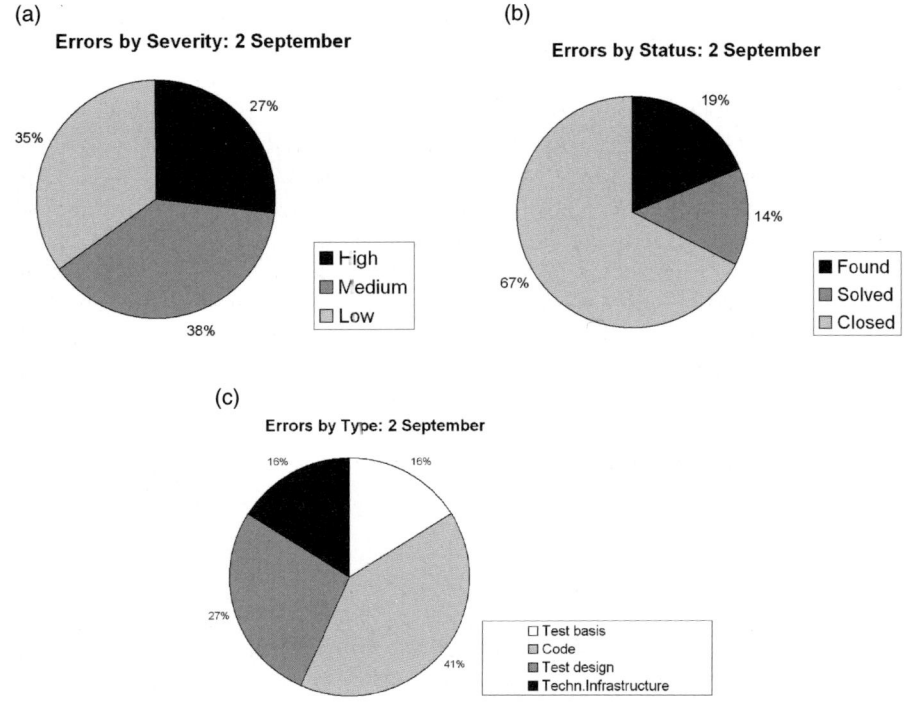

Fig. 20.6 Errors by severity, status and type

20.2.10 Test Result by Risk Category or Test Cluster

The below test result graphics display the test results by risk category or test cluster . A bar chart is use to indicate how many tests were successful or have failed. The graphic provides insight into the quality of the system and into the progress of the test project. The number of tests yet to be run indicates how the test is progressing.

Depending on the target group, the test result can be displayed by risk category or test cluster.

- **Representation by Risk Category**
  The test results graphic by risk category can be used if the target group does not have any knowledge of or interest in the system's functions. This will be the case for reports sent to upper management, the program manager or the business manager. If management is more interested in the risks than in the actual workings of the system, it does not make much sense to indicate in which remote part of the system an error was found. What they do want to know is

whether the most important risks have been covered, which is why a report by risk category is appropriate.

- **Representation by Test Cluster**
  If a risk analysis was not done, reporting is done by system function or test cluster. This graphic is suitable for those who are involved in the content, such as a technical project leader who knows the content of the clusters and knows which programmers are responsible. He is also often aware of each of the cluster's unsolved errors and change requests. The tasks are often planned by cluster or function. This graphic ties in better with the daily experience of those involved. The graphic indicates precisely how many tests were run for each cluster and what the quality of the cluster is.

Below is a profile of both kinds of graphics.

Test Result by Risk Category

The test result by risk category graphic provides an overview of the results of the tests that were run up to a specific point in time for each risk category defined in the risk analysis. The graphic in Fig. 20.7 shows how many tests were successful, how many failed and how many still have to be run.

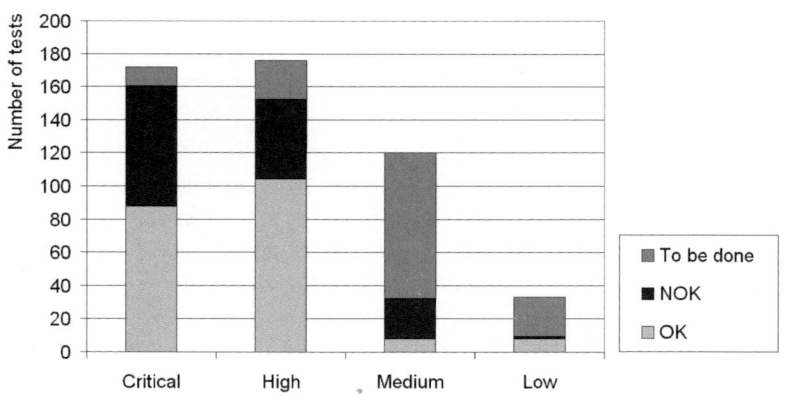

Fig. 20.7 Test result by risk category

This graphic indicates that most of the tests have been run for the two highest risk categories. It looks like the risk was taken into account when determining the sequence of the tests and that the tests in low risk categories were assigned a low priority.

The graphic shows that about half of the tests have failed. A test fails when the system behaves differently than expected. This is the case if the system contains a bug, if the test was not run correctly or if the test case was not correct. In the first case, the found error should be solved. In the other cases, the test set will have to be modified. After the error has been solved, a retest will be run to see whether the test is successful.

It is interesting to examine why some tests did not run. Tests are often stopped by showstoppers, but can also stop because they are simply not feasible. If a test is not feasible, determine whether the test is important or not. If it is not important, remove it from the test set. If it is important, the fact that it has not run is an outstanding risk.

Test Result by Cluster

The test result by cluster graphic (Fig. 20.8) provides insight into the number of tests defined for each cluster, and into the test results for each cluster. The clusters in this graphic correspond to the areas of attention defined in the TRA.

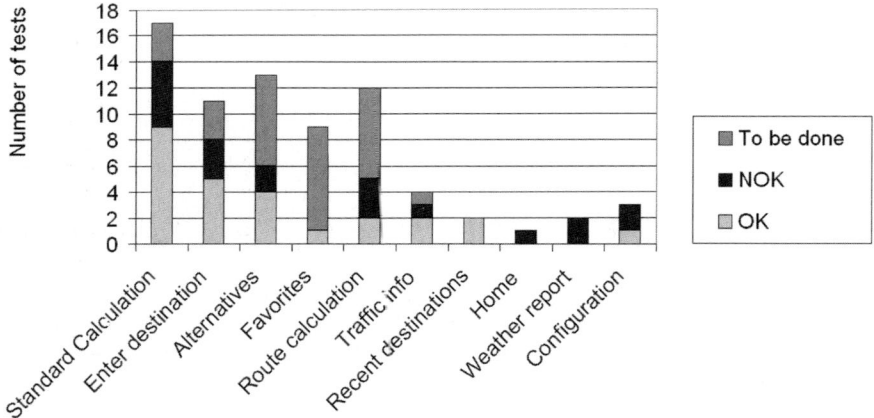

Fig. 20.8 Test result by cluster

Noteworthy is that more tests are defined for some clusters than for others. This difference is due to the size of the test cluster and the variation in test coverage, which is determined by the clusters' relative importance. Important clusters are tested thoroughly and with heavy techniques and contain more test cases. The functions with fewer risks are tested less thoroughly and contain fewer test cases.

Depending on the target group, the graphic can also display the test results by function or quality attribute. By using percentages rather than the absolute number of tests, the difference between the number of tests for each risk area or cluster is filtered out of the report.

When a fourth status is used, the statuses become OK, NOK, To be run and Not feasible. A disadvantage of this additional status is that a test that is not feasible is quickly discarded. The report must clearly state whether the test is important or not. This is why only three statuses are used in the example in Fig. 20.8.

20.2.11 Test Progress – Executed Versus Planned Tests

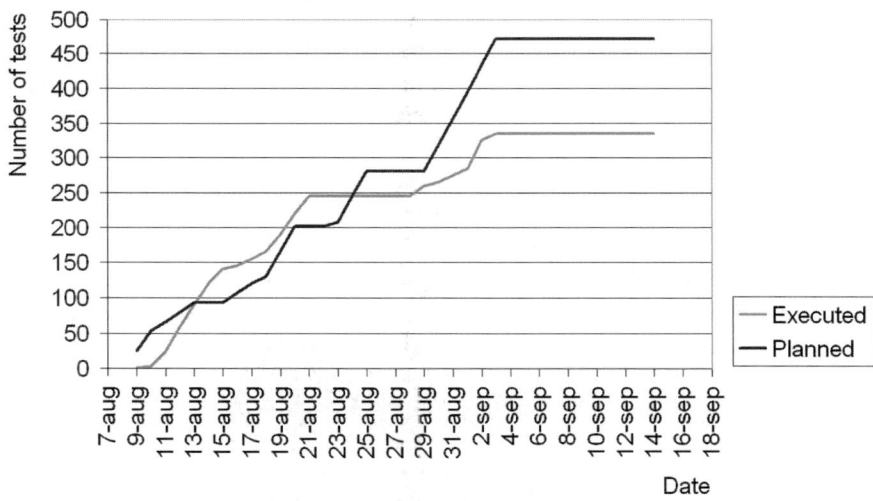

Fig. 20.9 Executed versus planned tests

In the graphic in Fig. 20.9, the number of tests run is compared to the goal. The counting unit will be different for each project and can consist of the number of unit tests, the number of check marks (OK/NOK) or the number of scenarios. For exploratory tests, this is the number of completed test charters.

The graphic contains two curves.

Planned
This curve shows the total number of tests that should be completed at a certain point in time according to the planning.

The curve is calculated using the planned available resources and the planned execution speed (total number of tests/total number of available resources).

The line is not straight because it depends on the number of available test hours in each period; for example, fewer tests are run during vacation periods.

The graphic in Fig. 20.9 shows that there are periods in which few resources are planned. The curve of the planned tests is horizontal. All of the tests are planned to be run on September 3, at which time the line is horizontal.

Executed
The curve displays the total number of tests that have actually been run regardless of the result.

When studying the curve of the number of tests that were run, it is notable that the test was behind target at the beginning of the project. Once it got going, the delay was made up for. Hardly any tests were run from August 21 onwards, which again put the project behind target. It is interesting to examine why this happened and if the error status graphic shows a similar trend.

There are a number of reasons why the "executed tests" line is horizontal.

- Fewer resources than expected are available
- Other tests are being run than those included in the planning; for example, errors were retested or more testing was done on problematic modules outside of the test design
- Testing cannot be continued because showstoppers have been found or because the test environment is not available
- The planning is not feasible

The graphic can also show the results by risk category instead of the total number of tests. The figure also shows that tests have been run, but it does not show whether they are the most important tests. This can be deduced from the "Test result by risk category" graphic. The progress can also be displayed in percentages instead of in absolute numbers.

20.2.12 Outstanding Product Risks

The "Outstanding product risks" report indicates whether each of the identified product risks has been covered by testing. The decision of going live with the system will mainly be based on the outstanding risks. To cover the most important risks, tests were run that enable the tester to make a statement about the degree to which the risks are real.

The degree to which the risks are a potential danger can be deduced from the following information:

- The tests that have not been run (yet) (see Sect. 20.2.6 Executed versus planned tests)
- The quality gap (see Sect. 20.2.7 Error status)

The outstanding risks can be reported in a textual overview or a graphic as shown in Fig. 20.10.

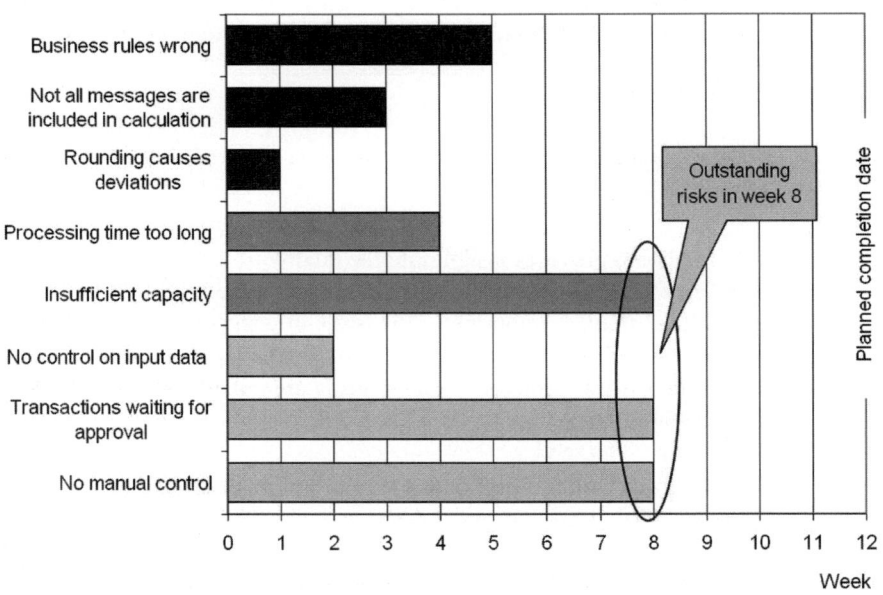

Fig. 20.10 Outstanding product risks by week

This graphic displays the risks that were identified during the 2D TRA (see the example in Sect. 7.3). At the end of each week, it is indicated whether the risk is still outstanding or whether it has been covered by the tests. The graphic shows that all of the product risks were still outstanding at the end of week 1. At the end of week 8, the number of outstanding risks was reduced to 3, and the other risks were categorized as no longer real. The tests either proved that the risks were not real, or the source of the risk was fixed. Critical errors that were found during testing were solved and retested.

Using this graphic has a number of advantages. The graphic illustrates that testing provides clear insight into the outstanding risks and thus adds value. Moreover, the graphic provides an instantaneous overview of the risks that were covered and that are still outstanding. Management can use this graphic to determine whether it is worth continuing the tests or if the indicated risks should be accepted and the system go live [Gerrard], [ISEB practitioner, 2004], [Gardiner, 2006].

20.3   The Dashboard

Including all of the reporting elements and an explanation in the test report may make the test report too lengthy and take too long to write. Consider using a dashboard in such situations [Clermont, 2006].

> A dashboard is a clearly structured compilation of up-to-date control data.

The dashboard provides insight into the quality of the test object and the progress of the test project. The stakeholders have specified the information they think is important. This is the information they need to determine the degree to which the anticipated goal has been achieved and to control the test project. These controls are called Key Performance Indicators (KPIs).

The dashboard must be updated regularly, for example, in real time [Koomen et al, 2007]. In many projects, however, it is sufficient to update the information once a day. Whatever you do, make sure you indicate when the dashboard was last updated.

The dashboard should be in a central location and can consist, for example, of a printout that is stuck to the water cooler, but it can also be published on the Intranet or in a wiki (see Fig. 2.2). This last option is especially effective when the work is done at more than one location. The dashboard in Fig. 20.11 uses five elements that were discussed earlier in this chapter. These elements provide insight into the quality, the progress and the costs of the test project.

Step 5 – Execution

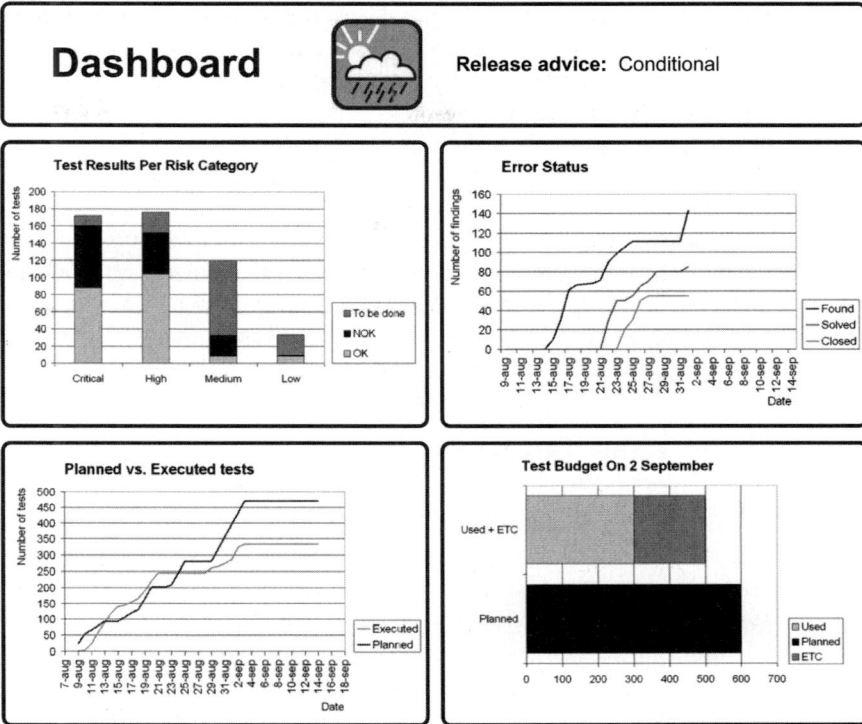

Fig. 20.11 A dashboard

The dashboard consists of the following elements:

Release Advice
Advice about the release moving to the next phase or going live. At the beginning of the project the advice will be "Negative." As soon as the tests demonstrate that the quality is good enough, the advice will change to Conditional or Positive.

Tests Run by Test Cluster, Risk Category, Function or Quality Attribute
The number of tests that finished with a negative or positive result is indicated for each risk category, test cluster , function or quality attribute, as is the number of tests that still need to be run. The graphic provides a quick impression of the observed quality. This report format depends on the target group of the report.

Error Status
The error status report indicates the number of critical errors that were found, solved and closed. The report provides insight into the quality of

the system and indicates whether the workload lies with the development team or the test team.

Executed Versus Planned Tests
This overview provides insight into the progress of the test run. Possible delays in the progress of the test will be clearly displayed.

Hour Estimate
The hour estimate overview indicates whether the project will be within budget or if it will exceed it. This helps control the financial aspects of the test project.

20.4   Clarity of the Test Report

A test project produces a lot of data. We need to know which data can help us formulate a clear message and support it. This is the point in time when data becomes information. As already mentioned, graphics are a good way of providing stakeholders insight into the status of the test project because they're a good way of visually displaying information. To ensure the message is understood, a number of issues need to be given attention:

1. Place Yourself in the Position of the Receiving Party

Remember that project managers receive a lot of progress reports at the end of the week. For this reason, progress reports should be simple and short, and contain a clear and understandable message. Irrelevant information muddles the message.

2. Take the Perceptions of Others into Account

Each party will be inclined to extract the information that confirms their perception. The tester will often be inclined to think in terms of errors. A report stating that 90% of the tests were successful will be interpreted by a tester as "10% of the tests were not successful." Project managers, on the other hand, will be very positive about "90% of the tests having been successful." This is why it's often necessary to add an explanation.

Example 20.3: The Connecta Dashboard

The status and the progress of Connecta's system test project are reported using a dashboard, which contains a limited number of graphics. To help the stakeholders interpret these graphics, David Bloom explained at the beginning of the project how he intends to report. He also indicates in the top right corner of the graphic [Mash, 2006] whether the graphic bears good or bad news. To do this, he established criteria together with the stakeholders.

Hour Estimate

 If the number of hours spent + ETC < 90% of the number of hours estimated.

 If the number of hours spent + ETC = 90-100% of the number of hours estimated.

 If the number of hours spent + ETC > 100% of the number of hours estimated.

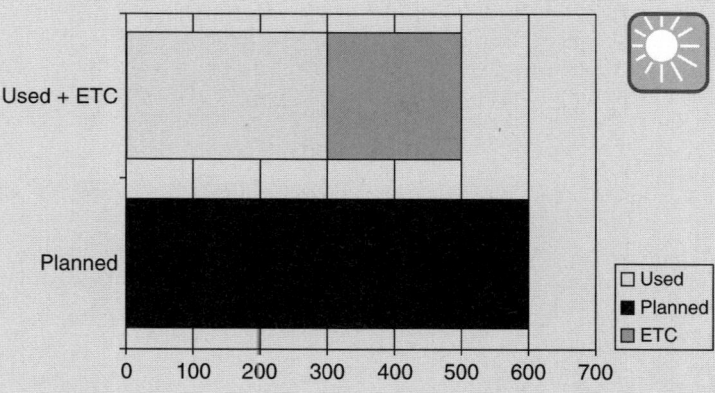

Fig. 20.12 The top bar is 83% of the bottom one. According to the criteria, the indication is positive (sun)

Test Results

Different criteria are used for the test results of the test project. It is expected that a number of tests will fail by the end of the first test run, which is OK. However, the criteria are higher for the regression tests, at the end of which all of the errors should be solved. This is why different criteria are used to determine the indicator during the regression test.

During the First Test Run

 If > 80% of all of the tests run is OK

 If 40-80% of all of the tests run is OK

 If < 40% of all of the tests run is OK

During the regression test

 If >90% of all of the tests run is OK

 If 80-90% of all of the tests run is OK

 If <80% of all of the tests run is OK

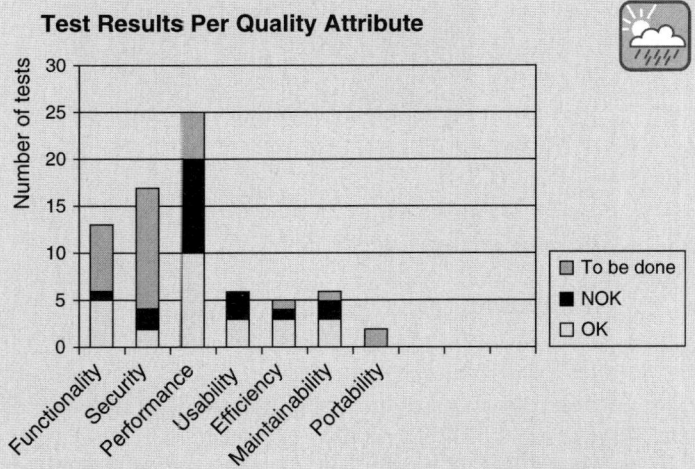

Fig. 20.13 79% of the tests have a positive result. For the first test run, this means that the indication will be neutral (cloud-sun)

Step 5 – Execution

The indication enables the stakeholders to determine the amount of attention they give to the graphic in question. They are also prevented from interpreting the graphic as positive while David is trying to issue a warning.

3. Include Underlying Data

To keep the report clear, the information is compiled and summarized. To make the report reliable, underlying data is included in the formulation of the message. Some specific problems will otherwise get buried.

Example 20.4: 90% was successful

In the meantime, the Connecta project has progressed quite nicely. During the system test, two test runs were completed and the development team has indicated that almost all of the errors have been solved. It's time for the final regression test. David reports the progress of the regression test: 90% of all of the tests were successful. Based on the graphic in Fig. 20.14, his test manager, Yasmin Hassouni, informs him that she is very happy with the quality of the system.

Test Results: 3 August

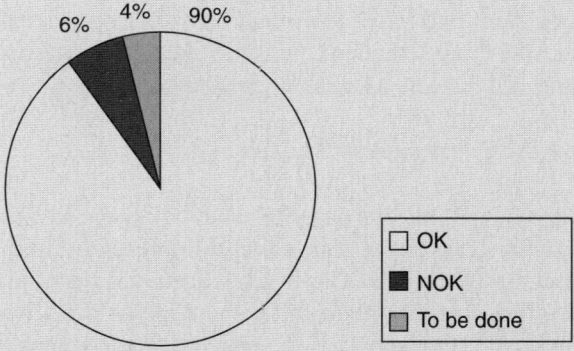

Fig. 20.14 The test results

David explains to Yasmin that appearance is deceptive: 6% of the tests were not successful. Drilling down into the test results by function shows that the failed tests can be traced back to one

Step 5 – Execution

function. The TRA shows that this function is crucial and that any related errors have a high severity. It is therefore safe to conclude that this function does not meet the requirements. Additional measures will have to be taken before a positive release advice can be given.

Test Results per Function

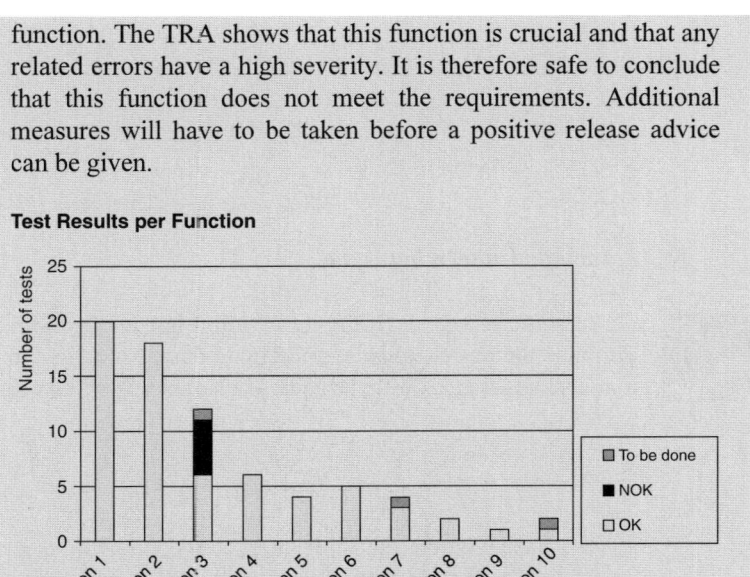

Fig. 20.15 The test results by function

4. Suggest Measures

Reporting that the test project deviates from plan is not a great achievement and adds little value. The test report increases in value if it also provides suggestions on identifying or eliminating the causes of the deviation. This makes the test report part of Deming's Plan-Do-Check-Act (PDCA) circle [Deming]. Include a concrete suggestion for possible measures and indicate who should implement them.

5. Have the Courage to Be Positive

A tester who only reports negative issues will soon lose the attention of his audience. Experience shows that a lot of testers find it difficult to point out positive things. This is often due to caution: "If I say the system is of good quality, they will come directly to my desk when they find an error." Everyone has to realize that a common goal is being pursued. This is why you should indicate what is going well and what needs attention. Indicate what management needs to focus on, but have the courage to give advice about the things that don't require much attention.

Step 6 – Assurance

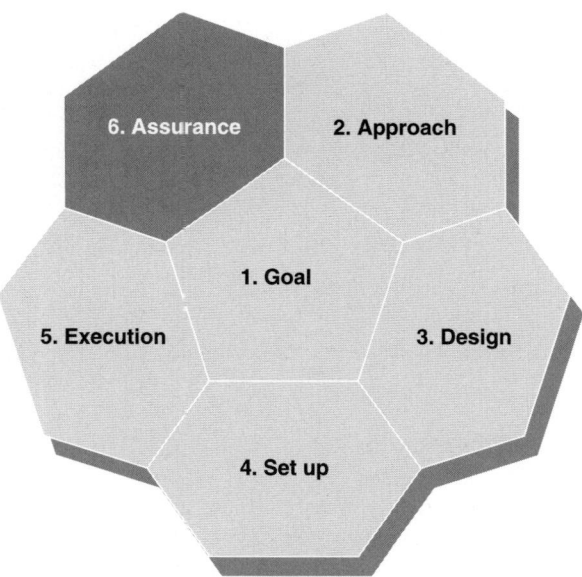

The Assurance step checks whether the test project's anticipated goal has been achieved. The organization of the test project is examined during the evaluation of the test project; what went well and which measures need to be taken to prevent the same problems occurring in the future. The test set is also evaluated to ensure it is reusable: superfluous tests are deleted and new tests added. The products that were used and developed during the test project are handed over, for example, by making the test set available to the maintenance organization. The handover ensures the receiving party knows how to use the products.

The Assurance step consists of the following activities and products:

Activity	Product
Evaluate the test project	Lessons learned report
Determine the regression test set	Regression test set
Clean up the archive	Updated test archive
Handover	
Discharge	

21 Assurance

21.1 Introduction

The test project is completed after all of the tests have been run and a positive release advice has been given. The test plan is often pulled out from the bottom of the pile because it contains most of the agreements that were made. The test plan defines the goal, the products that should have been created and to whom they have to be handed over. This chapter describes the final activities of the test project:

- Evaluating the test project
- Determining the regression test set
- Archiving and securing the testware
- Handover
- Discharging the test team

The project is only really finished after the team has been discharged and the testers can focus on their other activities or on new activities.

21.2 Evaluating the Test Project

21.2.1 Purpose of the Evaluation

In order to repeat successes and avoid mistakes in subsequent test projects, it is necessary to determine the lessons learned during a test project. This enables upcoming test projects to be run more efficiently.

The evaluation can be part of the project evaluation, which is conducted according to the applicable project management methodology. If it's

D.-J. de Grood, *TestGoal*,
DOI: 10.1007/978-3-540-78828-7_21, © Collis B.V., Leiden, The Netherlands, 2008

not, the test coordinator will do the evaluation himself by formulating a lessons learned report and handing it over to the right person. This can be the manager of the test department or the QA department, or the test manager who is responsible for the project's other test projects. Note that the test plan already determines how the handover is done.

During the evaluation of the test project, the actual results are compared to the anticipated goal. The effect that the expected and unexpected events have on the test projects is analyzed. This analysis is used to determine the successes and failures of the test project:

- Successes (with underlying events and experiences) are documented so they can be repeated
- Errors (with underlying events and experiences) are documented so they can be avoided

During the evaluation, the test coordinator gathers information about the events that occurred in the test project and asks the people involved about their experiences. Information can be gathered in a number of ways, for example, in group discussions or one-on-one interviews or by writing up a short overview. The method the test coordinator chooses is usually determined by the importance the project has for the organization. Because the staff often has to quickly move on to other test projects, there is usually not much time for the evaluation. It will be especially hard to find the time to evaluate less exciting projects.

The test coordinator adds his own observations to the information he has gathered and completes the lessons learned report, which he finalizes after it's been reviewed by the stakeholders. The lessons learned report is handed over to the person who is responsible for the next test project(s). The person accepts the lessons learned as advice and ensures it is taken to heart.

21.2.2   Points of Attention

In order to compare the results of the test project to the original objectives, the release advice and test reports are compared to the master test plan and the detailed test plan. The actual results are compared to the original plans.

To ensure that relevant events and experiences are revealed during the evaluation, the tester takes a look at a number of aspects:

Project Management

How was the test project embedded in the project (in terms of control, management and content)? Was quality given enough attention? How were relations between the project manager and the test coordinator and between the project manager and the test team?

Creation of the Product to Be Tested

How were the requirements, release and configuration management for the project set up? Which elements were set up well and contributed to a smooth test process? Which elements were missing and caused confusion and disrupted the test project? How was the quality of the deliverable (product and release notes)? Which deliverables were useful and which deliverables were poor? What were the consequences?

Test Strategy

To which extent did the test project enable the choices made in the test strategy to be carried out? What are the most important reasons for possible deviations? And why was the course not adjusted or the strategy adapted in the meantime?

Test Process

What phases was the testing divided in to? How did the preparation, the execution and the completion go? How were exit and entry criteria handled between the different process steps? Which entry and exit criteria were ignored, and what were the consequences? How was the release advice formulated, and the how were the decisions about the release and the handover made?

Test Team

What was the composition of the test team in terms of knowledge and experience, and in terms of their personalities and roles? Were system developers, administrators and users sufficiently involved? What was the atmosphere in the team? Why was this good or why does it need improving? How were relations between the test coordinator and his team? Were enough resources available, both in terms of quantity and quality?

Test Methods and Techniques

Which methods and techniques were chosen? Did they correspond to the choices in the master and detailed test plans? If they did, did the choices make the expected contribution? If they didn't, why was the plan not followed and why was the impact positive-negative?

Supporting Resources (Including Tools)

Were tools and other resources used? If they were, what are the experiences and how did the tools and resources contribute to the result and the goals? Were enough test environments available (quality and quantity)?

Step 6 – Assurance

21.2.3   Lessons Learned Report

The evaluation produces a lessons learned report. The findings are gathered for each finding, and for each finding the lessons learned are indicated.

Below are a number of examples:

No.	Finding	☺ or ☹	Lesson learned
1	Some test cases require a lot of preparation before they can be run.	☹	The current test design clearly indicates what the preconditions are. These preconditions have to be created manually. Preparing the operational data can save time during the test run, but the test data needs to be maintained properly. *Improvement suggestion*: In subsequent test projects, schedule time to set up good test data management.
2	The ad-hoc knowledge transfer between the analysts and the test team was experienced as useful. The test team gained a better understanding of what it had to pay attention to, which improved the quality of the defined tests.	☺	Knowledge about the system that is to be tested is needed to design the tests. *Improvement suggestion*: Organize knowledge transfer sessions with analysts and senior developers at the beginning of *every* test project.
3	Using the X-simulator considerably reduced testing time.	☺	Good use of tools improves the test efficiency. *Improvement suggestion*: Investigate whether the other interfaces can also be stimulated.
4	Deployment produced a lot of problems. The test had to be interrupted several times because the test object was not properly configured. The wrong modules were delivered more than once.	☹	Configuration management needs to be improved. *Improvement suggestion*: Determine whether the configuration is OK during the smoke test. Testing will not start if this is not the case.

21.3   Determining the Regression Test Set

The regression test set is one of the products that were defined in the test plan as a project deliverable because the organization expects it will need the tests again.

In a software development project in which a new system is built, the new system will have to be maintained after it has gone live. Maintenance is necessary to fix bugs or make changes to the functionality. If the test object is approved during the conformity, interoperability and certification tests, the object will not be resubmitted to this test set in the near future. This doesn't mean that the test set cannot be used to certify other test objects that have to meet the same requirements. This is a good reason to assure the tests can be reused.

Flaws in the test design will be discovered during the test run; interpretation errors of the tester, an assumption that wasn't entirely correct, or a test base that was not tested after it was changed.

Experience shows that some of the test cases do not add value. Although the tests are correct, it's better to remove them from the test design. An example is a function that uses generic code. If the testers do not know that the function is based on generic code, more than one test case will be designed to test the same code. In the regression test, one test case will then be enough. Another example is a test whose results are unaffected by change. This can be the case if the test did not produce any errors during the entire test project, even though a lot of releases were tested. This test will not have to be run in every regression test.

The purpose of the activity test is to clean up the existing tests. The result is a test set that indicates which tests are important for retesting. The tests may be accompanied by an explanation. It is advisable to specify the risk category a test belongs to the test design. This makes it possible to distinguish between full regression tests and, for example, quick scans. Indicating the importance of a test action enables the regression test policy to be easily defined, as shown in the below table (see also paragraph 13.4 Test actions):

Type of regression test	Execute
Quick scan	Only the tests in the risk category "critical"
Regression test	Only the tests in the risk category "critical" and "high"
Full regression test	All tests

Step 6 – Assurance

Admittedly, determining the regression test set is not the most exciting activity. It can even be hard to remember the whole process and all of its details at the end of a project. However, it is important to do it fairly soon after the completion of a project because it will only get harder as time goes by.

21.4   Archiving and Securing the Testware

Not only the tests need to be secured, the testware does too. Once the test project is finished, all of the products and tools need to be findable and reusable, as do the manuals for the test tools. If the project is run in a controlled way, many of these things will have been taken care of already. If not much attention has been paid to this, it is advisable to take care of these things before the end of the test project. If there is no time to document everything, it is definitely useful to document which project members have knowledge of the testware.

The error logs, relevant e-mails and memos describing the agreements also have to be secured. The testware is secured by saving it in the archived project directories or on an Intranet. Each organization will have its own procedures.

21.5   Handover

In the test plan, we already defined to whom we would transfer what knowledge. Now it is time to put these plans into effect. This activity can be easily combined with the securing of the testware. The receiving party often has its own wishes about how and where they want to store the products.

Handover means that the products are not simply "thrown over the fence," but that it is checked whether the products have been accepted. The receiving party also needs to take on the knowledge. In a good handover, the receiving party determines whether sufficient knowledge and material have been transferred.

21.6   Discharging the Test Team

The achieved result has been evaluated and compared to the anticipated goal. The products and knowledge have been secured and handed over. The project has been completed and the test team can ask the customer to discharge them. Discharge means that the test team members are dismissed from their responsibilities.

Experience shows that customers often hesitate when it comes to writing a formal discharge. This may be due to a fear that they will have no one to turn to when something goes wrong. This is a pity, and would not be necessary if insight were provided during the test project into the working method, the quality of the products and the risks.

If a formal discharge cannot be obtained, make sure you complete the project anyway. Celebrate the success with a meal or a glass of champagne and share the success with the rest of the organization.

May your test projects be successful!

Appendix A – Checklist: Sanity Check on the Design

A.1   Conclusion

Conclusion	Y / N	
The specifications (test base) describe the system clearly and precisely. They are therefore sufficient as the basis for a structured test project.		

If the conclusion is negative:

High-level errors that were recorded during the sanity check are displayed below. For each error, indicate the risks for the test design and the measures needed to rectify them.

Error	Risk	Measure

A.2   Result

Description	Y/N	Not applicable	Solution
The functionality, the processes and their coherence is described sufficiently			
The functionality, the processes and their coherence concur with the anticipated goal			
All of the relevant quality requirements have been sufficiently taken care of			
The result of the risk analysis can be traced in the test base			

A.3   Control

Description	Y/N	Not applicable	Solution
Contains history log (including version administration)			
Contains approval/distribution list			
Contains document description (including author)			
The document has a status			
Contains reference list			

A.4   Structure

Description	Y/N	Not applicable	Solution
Contains executive (or management) summary			
Contains table of contents (current)			
Contains introduction			
Contains chapter(s) with an overview or description of the functionality (and related screens/reports and messages) Contains chapters with data models and data descriptions			
Contains chapters with functional specifications			
Contains chapters with non-functional specifications			
Contains chapters with interface specifications			

A.5   Content

Concerned functionality	Y/N	Not applicable	Solution
Contains system overview • System description • Purpose of the system, what is the use for the users and how does it support their business processes (high level) • Scope diagram/system diagram • System environment and dependencies • System owners, both functional and technical • Sites where the system will be implemented The system contains various subsystems • For every subsystem: a short description with basic functionality, processes and related data • Boundaries of and between subsystems • Relations between subsystems Quality attributes • Quality attributes are defined in the functional descriptions • Relevant quality attributes (e. g. performance, security) are defined for the whole system. • Relevant quality attributes (e. g. performance, security) are defined for each subsystem • Quality attributes are defined in functional descriptions.			
• Contains workflow/process flow/state transition or other diagrams • Description • Lists actions, conditions and situations • Describes actions, conditions and situations			
Contains use case (UC) • Use case description • Use case diagram (UCD) • Connection between UCD and relevant business processes • UCD contains boundaries • UCD contains primary and secondary UC			

Concerned functionality	Y/N	Not applicable	Solution
• Connection between UC and functionality • UC contains primary scenario, alternative scenario and fallback scenario			
Contains screen information • Screen layout • Screen/system navigation • I/O information			
Contains report information • Report description • Parameters • Fields • Grouping and sorting • Report layout • Contents			

Data model and data description	Y/N	Not applicable	Solution
Contains graphic model (ERD)			
Contains description of the entities • Entity definition: a clear and unambiguous definition of the notion of entity • Attribute definition: a clear and unambiguous definition of every attribute of the entity, including a reference to a domain or description of a type of data, length and permitted values • Primary and foreign keys: the attributes that identify the (related) entities • Quantities: minimum, average and maximum number of entities and growth rate per period			
Contains description of relations • Name: of the relation and the inverse relation • Cardinality: how many entities are linked to the relations • Option: the extent to which the relation is mandatory or optional for the entities of the relation • Constraints: have the regulations been established that determine how the system should react according to the cardinality and possibilities within the relations?			

Data model and data description	Y/N	Not applicable	Solution
Contains description of domains • Name • Data type and length • Permitted values			
Contains description of constraints			
Contains description of regulations			
Contains description of stored procedures			
Contains description of triggers			
Contains description of defaults			

System interfaces	Y/N	Not applicable	Solution
Contains the name of the interface; for example <interface System A – System B>			
Contains description of the purpose			
Contains workflow/scheme/diagram			
Contains details about the process			
Contains description of the type of interface: operational/data/batch/online			
Contains detailed information about the format of the interface: structure, data elements, data types, length, permitted values			
Contains description of error handling			
Contains description of error logging			
Contains description of constraints			

Functional specifications	Y/N	Not applicable	Solution
Contains name of the function			
Contains purpose of the function			
Contains process information			
Contains I/O information			
Contains description of data processing			
Contains description of calculation functionality			
Contains description of constraints (input checks)			
Contains description of error handling			

Authorization	Y/N	Not applicable	Solution
Description of the required authorization (profiles)			
Contains authorization matrix			

Technical architecture	Y/N	Not applicable	Solution
Description and/or schemes of the infrastructure, for example, network, servers, DBMS, operating systems, middleware			
Description of technical aspects of interfaces			

Appendix B – Checklist: Sanity Check on the Testware

B.1   Conclusion

Conclusion	Y / N	
The testware is described accurately enough and/or works well enough for reuse		

If the conclusion is negative

High-level errors that were recorded during the sanity check are displayed below. For each error, indicate the risks for the test design and the measures needed to rectify them.

Error	Risk	Measure

B.2   Result

Description	Y/N	Not applicable	Solutions
The functionality that is to be tested and the processes and their coherence are sufficiently described			
The functionality and the processes and their coherence concur with the anticipated goal			
All of the relevant quality attributes have been sufficiently taken care of			
The outcome of the risk analysis is traceable in the test design			
Test results from the last time the testware was used are available and provide insight into the quality of the tests			
Errors that were found after last time the testware was used are available and indicate on what points the coverage of the testware is insufficient			

B.3   Logical tests

Description	Y/N	Not applicable	Solution
The test basis on which the logical test is based is clearly specified			
The logical test design is up to date, there are no major outstanding issues in the design specified in the test base			
The structure of the test design is clear			
Logical tests can be traced back to related specification elements so changes in the test base can efficiently implemented			
Logical tests can be traced back to the related risk category in the TRA so changes in the TRA can be efficiently implemented			
Logical tests are set up on the basis of test techniques			
Where needed, test techniques are applied correctly and completely			
Choices that have been made in the logical test design have been clearly indicated (for example, deviations from the technique)			

B.4 Physical test

Description	Y/N	Not applicable	Solution
The test base on which the physical test case is based is clearly specified			
The physical test design is up to date, there are no major outstanding issues in the design specified in the test base			
The structure of the test design is clear			
The physical tests can be traced back to the related specification elements so changes in the test base can be efficiently implemented			
Physical tests can be traced back to the related risk category in the TRA so changes in the TRA can be efficiently implemented			
Physical tests can be traced back to the related logical test design so changes in the logical test design can efficiently implemented			

B.5 Test cases

Description	Y/N	Not applicable	Solution
Test cases have a unique ID			
Test cases contain all the required fields			
Test cases are detailed, it is clear what the purpose of the test is and when the test is successful			
Test cases clearly describe which test actions will be executed			
Test cases describe the input data that is used and the expected outcome			

B.6   Test tools

Description	Y/N	Not applicable	Solution
The test base on which the tooling is based is clearly indicated			
The tooling is up to date, there are no major outstanding issues in the design specified in the test base			
There are errors in the tooling and they have a clear status. It can be estimated which bug fixes/ changes will have to be performed on the tools			
Documentation for the tooling is available, so changes can be efficiently implemented			
The infrastructure is such that the tooling can be tested before the actual test run starts			

Appendix C – Checklist: Checklist smoke test system

C.1   Conclusion

Conclusion	Y / N
The system can enter the p anned test phase. It is expected that the tests will run without too many obstacles or problems	
The required products are available and their content is sufficient	
The system is sufficiently stable and has enough functionality to move on to the test phase	

If the conclusion is negative

High-level errors that were recorded during the sanity check are displayed below. For each error, indicate the risks for the test design and the measures needed to rectify them.

Error	Risk	Measure

C.2 Delivered products

Description	Y/N	Not applicable	Solution
Release notes are available and contain • The version of the release • Changes in relation to the previous release • The errors that were fixed • The implemented change requests (RfCs) • The known errors that can impact the test process			
Installation instructions			
User guide			
Test report for previous test phase			
Release advice for previous test phase			

C.3 General

Description	Y/N	Not applicable	Solution
The application can be started			
The user can access the application			
The user can perform elementary navigation without the application crashing			
The new functions are available and accessible			
It is possible to request, change or delete information (CRUD actions)			
The user can navigate from the application to interfacing systems without it crashing			

C.4 QuickScan

Description	Y/N	Not applicable	Solution
The most important correct path (feasible path) can be fully executed or The (automated) QuickScan can be performed without major problems			

Appendix D – Checklist:
Test charter exploratory testing

Charter ID	
Priority	
Time available (h)	

Conclusion	Y / N	
This charter has been completed		
Extensions of this charter will not be needed in a next session		
The anticipated goal of this charter has been achieved, the risk is sufficiently covered		
As far as this charter is concerned, can we go live?		

If the conclusion is negative

Indicate which follow-up action is needed to achieve the anticipated goal of this charter. Indicate which risks are not sufficiently covered and if necessary make suggestions for follow-up charters.

Error	Risk	Activity

D.1 Preparation

Anticipated goal of this charter	Priority

Exclude from this charter	Motivation

Why test this charter	

Expected problems	

D.2   Test log

ID	Test action(s)	Expected system reaction	Conclusion

D.3   Errors

ID	Description	Severity	Follow up action

Appendix E – Glossary

1D TRA

Test risk analysis during which the relative priority of functions and areas that need attention are estimated. The result in a one-dimensional risk matrix, i. e. a list of the risk areas that is sorted by relative importance.

2D TRA

For 2D TRA, the chance and the impact are estimated separately for each of the areas that need attention. The risk matrix displays the risks.

Acceptance criteria

The exit criteria that a component or system must satisfy in order to be accepted by a user, customer, or other authorized entity [ISTQB, 2007]

Action word driven

Test automation strategy that uses action words.

Anticipated goal

The goal that has been set out to be achieved. In Test-Goal, this is the result that supports the organization's goals.

Baseline

The state of a configuration, process or set of system specifications that is established at a specific time. For system specifications, this is a controlled set that is used as a basis for testing and system development.

Black-box test

A test that does not check the code and tests the system from the outside. The tests are generally run using existing system interfaces.

Boundary value	The minimum or maximum value of an equivalence class or domain.
Bug	See Error
Build number	Version number of the test object.
Chain test	Test during which systems are linked to each other. Focuses on finding errors that occur when systems do not interact well. The chain test is also referred to as system integration test (SIT), end-to-end test, or "integration in de large."
Change	The addition, deletion or change of things, such as configuration items, processes and documentation that can impact IT services. Also referred to as Request for Change (RfC).
Change Advisory Board (CAB)	A group of people that help the change manager evaluate, plan and prioritize a change. The CAB is usually made up of business representatives and third parties such as suppliers [ITIL, 2006].
Change Control Board	Consultative body that decides which changes will be implemented in the system.
Compliance	The degree to which a process or system meets a standard or a guideline.
Concurrency	A measure for the number of users that can perform an operation at the same time [ITIL, 2006].
Configuration data	Data that determines the configuration of the system. A requirement of testing is that the system is well configured. That's why it's important to have the right configuration data, which can be different at each test level.
Configuration management	The process that ensures that the configuration of the test environment is controlled and known.
Configuration/release version	The composition of a component or system as defined by the number, nature, and interconnections of its constituent parts [ISTQB, 2007].

Configure

Set up software or make changes to the settings of test equipment or the operating system/software [SurfNet]

Conformance testing

The process of testing to determine the compliance of the component or system, also known as regulation, compliance or standards testing.

Coverage

See test coverage.

Cross-check matrix

Matrix that specifies which physical test case is used in which test scenario.

Dashboard

A dashboard is a structured collection of current performance indicators, for example, an online collection of graphics that provide insight into the progress of the test project and the quality of the test object.

Data cycle test

Test technique that enables data to be added, deleted, changed or displayed. This technique is also referred to as CRUD test.

Data driven

In the data-driven approach, the test data is separated from the scripts. We call this parameterizing.

Defect

A flaw in a component or system that can cause the component or system to fail to perform its required function, e. g. an incorrect statement or data definition. A defect, if encountered during execution, may cause a failure of the component or system [ISTQB, 2007], also refered to as Error

Defect detection rate

Ratio of the number of reported errors for each time unit of testing.

Deliverable

Any (work) product that must be delivered to someone other than the (work) product's author [ISTQB, 2007].

Deployment

Process that makes the code of the test object available in the test or live environment. Also referred to as promotion, installation or roll-out.

Deployment cycle	Speed at which deployments can be carried out. A fast deployment cycle means that new system versions are deployed at short intervals. The time between which an error is found and the solution is available can be short. An example of a very short deployment cycle is a developer who runs the test, corrects the errors and reruns the test himself. In long deployment cycles, more time is needed before bug fixes can be tested.
Depth	See test depth.
Detailed test plan (DTP)	Approach for a test project. Also known als level testplan or phase testplan indicating that the plan describes one test level or one testphase [ISTQB, 2007]
Diacritical characters	Diacritical characters are characters that have an accent. The accent can be placed above, below or even through the letter to change its pronunciation. For example: â, é, ë, ç, ħ, ø, ß, æ. Diacritical characters are frequently displayed in a special way in the character set. In an HTML file, an *à* is entered as *&agrave* [Wikipedia].
Discharge	Release someone of their duties. Discharging the team releases it of its duties and responsibilities.
Domains	An area, field: an area that is covered by a test technique.
Drill down	Zoom into the underlying details of general information. For example, investigating the total test result for a system (general). Drilling down means looking at the test result for each function, quality attribute or risk category (underlying information) [Webopedia].
Driver, stub	Pieces of code that replace absent program code.
Dynamic test tools	Dynamic test tools are used for dynamic tests. These are the tests whereby the system is really used.
Effective productivity	The period during which an employee is really effective. An employee is effective for 60 to 80% of his work time.

End-user   See user.

Equivalence class   A portion of an input or output domain for which the
                            behavior of a component or system is assumed to be
                            the same, based on the specification [ISTQB, 2007].

Error   Deviation of the component or system from its ex-
                            pected delivery, service or result due to a human
                            action that produces an incorrect result, also known
                            as a failure. Free after [ISTQB, 2007].

Error analysis   An (statistical) analysis method that provides infor-
                            mation about the reliability of the measurement result
                            or the result of a calculation.

Error guessing   A test design technique where the experience of the
                            tester is used to anticipate what defects might be
                            present in the component or system under test as a
                            result of errors made, and to design tests specifically
                            to expose them [ISTQB, 2007]

Error management   The process of recognizing, investigating, taking
                            action and disposing of errors. It involves recording
                            errors, classifying them and identifying the impact
                            After [ISTQB, 2007]. Also known as Defect or Prob-
                            lem management.

Estimate to Complete   An estimate of the total number of hours that are
(ETC)   needed to achieve the agreed result. In contrast to the
                            estimated date of completion, the ETC doesn't con-
                            tain any information about the date on which the task
                            will be completed.

Estimated Date of   The date on which the agreed task is expected to be
Completion (EDC)   completed. In contrast to the estimate to complete, the
                            EDC doesn't contain any information about the effort
                            required to complete the task.

Exit and entry criteria   The criteria that have to be met before a process can
                            start or stop.

Failure   See Error

Fault   See Defect

Feasible path	Sequence of valid tests. A path for which a set of input values and preconditions exists which causes it to be executed [ISTQB, 2007]. The tests focus on how the system behaves under normal use and do not take the system's error handling into account.
Finding	Findings indicate an observed difference between expected and implemented system behavior that can jeopardize the anticipated goal. This definition includes both the experience of the tester and the anticipated business goal. A finding can originate from a test error, an error in the test base, or an error in the code.
Fit for purpose	Degree to which the system does what it's supposed to do.
Fixed price	Assignment for which the agreed result is delivered at a previously agreed price.
Formal reviews and inspections	Formal inspections carried out to find errors in the documentation. A formal inspection is a technique whereby the reviewers are trained for the process and work according to clear procedures [Wikipedia].
Functional acceptance test (FAT)	The functional acceptance test checks the functional operation of the test object against the system requirements and the functional design.
Functional decomposition	The decomposition of a system into its individual functions. The functional decomposition can be used in the test rick analysis and is often displayed as a test tree. The functional decomposition can contain quality criteria in addition to functions.
Functional design	A detailed description of the user and other specifications of the information system or of the changes that will be made to it so that they can be implemented and tested unambiguously [ASL].
Functionality	The capability of the software product to provide functions which meet stated and implied needs when the software is used under specified conditions [ISTQB, 2007].

Generic test strategy	Generic test strategy that is created for several master projects or that specifies how tests should be run in one or more organizational divisions.
Goal description	Formulation of the assignment in which the anticipated goal of the test project is defined.
Heuristic	The methodical way of learning by trial and error [van Dale (Dutch dictionary)]. This includes using checklists to reach a conclusion.
Impact	A measure of the effect that an incident, problem or risk has on the business or other processes. The impact and the urgency determine the priority.
Improvement report	Project evaluation report. The improvement report contains the lessons learned during the project.
Input and output data	Data is input during the test and the result of the test action compared to the expected result. Input data can be seen as data that is used to determine boundary values, in syntax tests, etc. Output data can be, for example, the results of a calculation, or it can consist of data elements that are generated by the system for messages or reports.
Installation or deployment manual	The installation or deployment manual contains instructions for the system administrator. This manual can be a checklist and defines which installations have to be carried out. Deployment can be very complex on large systems, which is why the sequence in which the installations, upgrades, patches and controls are carried out must be clearly described.
Interoperability	Exchangeability of the test object in relation to other systems.
Iterative process	Process that falls back on previous steps. Parts of the process are repeated one or more times.
Life cycle	The subsequent steps that a test project or software development project goes through. The life cycle of the test project is described in the TestGoal step plan.

	The steps in the V model provide a basic description of the software development cycle.
Live	A system that is actively being used to process business data.
Live environment	Hardware and software products installed at users' or customers' sites where the component or system under test will be used. The software may include operating systems, database management systems, and other applications [ISTQB, 2007]
Load	A measure for the number of concurrent transactions or users that a system can process.
Logical test design	See test design.
Maintenance	All of the tasks, responsibilities, and activities that are required to keep objects in such a state that they continue to meet the defined requirements and needs of their owners [ASL].
Master test plan (MTP)	Approach for a master test project.
Master test project	Type of structure in which test projects that belong together are grouped. The master test project contains all test levels.
Metadata	Predefined data that is needed to create a specific configuration. Metadata is also called master data.
Milestone	A measurable status or deliverable that has been predefined.
Moderator	Person who monitors and supervises a process and keeps it going. For example, a formal review or exploratory test debriefing.
Module test	Test that focus on the code's elementary building blocks or individual software components [ISTQB, 2007]. They show that the modules technically meet the system requirements. Also referred to as unit testing or component testing.

MTP	Master test plan, approach for the master test project.
New development	The making of a new application.
Null	In computer science, the term "null" is used to designate a missing or unknown value. Null values are used in a variety of programming languages and in databases [Wikipedia]. Null is used to specify that nothing is entered in a field, for example, to test that a form can only be saved if all of the fields have been filled in. The form cannot be saved if one of the mandatory fields contains a null value.
Operational data	Data or a process in a specific state that is present in the existing data set and meets the preconditions of a test case.
Organization chart	Graphical display of the organization. The organization chart displays the organizational divisions and their hierarchical relationship.
Parsing	The conversion of files or data to smaller or other files or data.
Peer review	Review by a colleague with experience with the corresponding topic.
Performance	The degree to which a system or component accomplishes its designated functions within given constraints regarding processing time and throughput rate [ISTQB, 2007].
Performance tests	The process of testing to determine the performance of a software Product [ISTQB, 2007].
Physical test cases	See test design.
Pilot	The pilot consists of testing the system in a simulated live environment prior to deploying new functionality. This ensures that errors that occur during the pilot will not occur in the live environment.
PKI	Public Key Infrastructure is a technology that plays a crucial role in the creation of reliable electronic ser-

vices. PKI ensures that electronic transactions are incontestable by placing a digital signature. PKI also ensures that information is safely sent across the Internet.

Pre-Condition   Attribute of the physical test design that describes the starting point of the test.

PRINCE2   PRINCE2 (PRojects IN a Controlled Environment) is a structured method for effective project management.

Priority   The value of an activity compared to other activities, which is expressed in the speed and/or thoroughness with which and activity needs to be carried out or completed compared to other activities [ASL]. The level of (business) importance assigned to an item, e. g. defect [ISTQB, 2007].

Product risks   Risks that jeopardize the anticipated goal. The product risks make a statement about the test object but not the process that creates them. Its a risk directly related to the test object [ISTQB, 2007].

Production acceptance test (PAT)   The production acceptance test focuses on releasing the system into the live environment and entering it into maintenance.

Production likeness Degree to which the test environment resembles the live environment.

Project   A project is a unique set of coordinated and controlled activities with start and finish dates undertaken to achieve an objective conforming to specific requirements, including the constraints of time, cost and resources [ISTQB, 2007]

Project brief   Document with general project information, such as start and end date, users/project members, project leader etc.

Quality   The degree to which a component, system or process meets specified requirements and/or user/customer needs and expectations [ISTQB, 2007].

Quality attribute

A quality attribute describes one aspect of the quality a system must have. Quality attributes create a conceptual framework that helps explicitly describe the desired quality.

Quality gap

The quality gap is the difference between the anticipated and the achieved quality. This difference is frequently measured according to the number of open errors. The quality gap is then the difference between the number of found and solved errors. In practice, the quality gap will never be closed. This means that there are risks for the live environment that have to be explained in progress meetings and in the release advice.

Quick scan

A fast and limited test where only tests in the risk category "critical" are run.

RACI chart

Matrix that contains responsibilities and involvement.

Record & Playback

Possibility to record user actions carried out on the test object in scripts and play them back, also known als capture & playback.

Regression test

Test that checks whether regression has occurred. Regression is the phenomenon whereby unchanged code no longer works as the result of changes made to the environment or underlying system components.

Regression test set

The test set that is used during the regression test.

Release

All of the new, changed or existing program items that are made available for exploitation at a given point in time.

Release advice

Advice to the customer concerning the test project's transition to the next planned phase. The advice is made based on the test results and the risks that are jeopardizing the anticipated goal.

Release management

The release management process ensures that changes are applied at specific times and it is known which changes will be applied.

Release note	The release notes describe the characteristics of the deliverable. They contain the version of the release, the changes made since the previous release, the solved errors, the implemented changes, and the known problems that can impact the test process, also known as item transmittal report.
Requirement	A formally defined condition that the system must fulfill [ASL].
Retest	Test that is run to determine whether a previously found problem has been correctly solved.
Review	The test technique whereby one or more people comments on a product. for the purpose of identifying defects and improvements. After [ISTQB, 2007].
Review logging form	Form in which the review findings are entered.
Risk category	Collection of risks that have the same priority.
Risk matrix	Graphical display of the risks in which the chance of each risk is set off against the impact.
Risk-based test	An approach to testing to reduce the level of product risks and inform stakeholders on their status, starting in the initial stages of a project. It involves the identification of product risks and their use in guiding the test process [ISTQB, 2007]
Sanity check	Activity that provides insight into the quality and testability of the test base. The sanity check specifies which measures need to be taken before a test object can be used without risks. In the Netherlands, sanity check is called "Test base intake."
Sanity check report	Result of the sanity test
Scripts	Program code that runs the tests. The scripts can be recorded with the automation tool's record and playback function or coded manually.
Security test	Tests that demonstrate that the system is sufficiently protected against potential abuse.

See test depth | Measure for the depth of the test technique.

Severity category | Values that describe the severity of an error. Commonly used values are: high, medium and low.

Significant | Non-negligible factor.

Simulator | The test tool communicates with the test object and simulates a user or the system.

Smoke test | Activity that provides insight into the quality and testability of the system. The smoke test specifies which measures need to be taken before a test can be started without risks. Also known as confidence test, intake test or sanity test.

Sox | The Sarbanes-Oxley Act was put in place to guarantee the integrity of annual reports. The Act was implemented in the United States in 2004. Companies listed on the New York stock exchange must show that they meet the requirements of the Act.

Stakeholder | Person or party that is interested in the quality of the test object. Stakeholders will want to influence the project or will want to receive information. They also provide the tester with important information about the information he has to include in his report.

Standard deviation | Term used in statistics to specify how measurement values are spread. For normal distributions, 68.2% of the measurement values are between the mean minus the standard deviation and the mean plus the standard deviation.

Static test tools | A tool that carries out static analysis on software artifacts, e.g. requirements or code, carried out without execution of these software artifacts. After [ISTQB, 2007].

Strong technique | Test technique with a wide coverage. The technique generates a large number of test cases.

Stubs, drivers | See drivers, Stubs

Success factors	Factors that determine the success of an activity or project.
Supporting test tools	In addition to tools that help test or analyze data, there are also tools that support the test process. Examples: • Error logging tools • Planning tools • Tools in which tests can be designed. • Tools in which test results can be reported (dashboard) • Configuration management tools
System	A collection of components organized to accomplish a specific function or set of functions. [ISTQB, 2007]. Example: A computer system with hardware, (embedded) software and applications.
System specifications	Description of the desired operation of the system.
System test	The process of testing an integrated system to verify that it meets specified requirements [ISTQB, 2007]. If a customer-supplier relationship exists, the system test is carried out by the supplier.
System under test (SUT)	See test object.
Technique matrix	Graphic in which test strength and domain coverage are displayed.
Test	The analysis of software items with the goal of finding errors and differences with the specifications [IEEE 829, 1998].
Test action	Concrete description of an action that a tester has to carry out when running a test case.
Test analyst	The analyst uses the test plan to determine which tests have to be run. The analyst creates the logical tests.
Test automation	The use of software to perform or support test activities, e. g. test management, test design, test execution and results checking [ISTQB, 2007].

Test automator	The team's automation expert.
Test base	All documents from which the requirements of a component or system can be inferred. The documentation on which the test cases are based. [ISEB]
Test charter	A statement of test objectives, and possibly test ideas about how to test. Test charters are used in exploratory testing [ISTQB, 2007].
Test cluster	Group of tests that belong together, for example, all of the tests that cover a specific area of attention, or tests that test the same function or subfunction. Test clustering determines the structure of the test tree. Each branch of the tree is a test cluster. The test cluster can often be recognized in the TRA.
Test coordinator	Tester who is responsible for the success for a test project.
Test coverage	The degree to which a specified coverage item has been exercised by a test suite [ISTQB, 2007]
Test data	Data that is used during the test run. TestGoal distinguishes four styles of test data, namely: Input and output data, operational data, configuration data, and metadata.
Test data repository	A test data repository is a central storage area for test data. Because all of the test data types (input and output data, operational data, configuration data and metadata) are strongly related to each other, it's efficient to manage them as one set. This is done in a central storage area, which can consist of a tool, a database or an Excel sheet.
Test depth	The test depth specifies how thoroughly each test case/risk area has to be tested.
Test design	The set of defined test cases. There is a logical and a physical test design. The logical test design describes *what* has to be tested. The physical test design describes *how* the test cases have to be run. The physical test design also contains test data.

Test design technique	Technique that defines how test cases have to be derived from the test base.
Test engineer	Tester who is responsible for the physical test design and the execution of the tests.
Test environment	Environment (hardware and middleware) that is used for the test system.
Test level	A group of test activities that are run and controlled together [Pol et al, 1999].
Test manager	Tester who is responsible for the success of the master test project.
Test object	The system being tested. The object that is being tested. The system for which a release advice is given. Also called "System Under Test" (SUT).
Test project	Type of organization in which test activities that belong together are grouped. A test project is often set up for each test level.
Test report	A snapshot that specifies what the achieved result is. This is done based on the progress of the test process and the quality of the test object.
Test Risk Analysis	The process of assessing identified risks to estimate their impact and probability of occurrence (likelihood) [ISTQB, 2007]
Test run	A collection of test activities whereby all of the tests have been run once. A test run can be conducted with several versions of the test object. At the end of a test run, all of the tests will have been carried out, but they may not all have been successful.
Test scenarios	Specifies the preferred sequence of the test cases. The sequence is often determined in such a way that connecting tests can be run back to back.
Test set	Collection of test cases, also known as test suite or test case suite.

Test strategy	The test strategy describes how the anticipated goal is translated to the way in which the tests are conducted. Among other things, the strategy describes the risk categories and prescribes the test techniques, also known as test approach.
Test strength	Each risk area is assigned a test strength, which is used to determine the test techniques.
Test tree	The decomposition of the test object into functions and areas of attention.
Testware	Artifacts produced during the test process required to plan, design, and execute tests, such as documentation, scripts, inputs, expected results, set-up and clear-up procedures, files, databases, environment, and any additional software or utilities used in testing. [ISTQB, 2007]
Time stamp	The registration of the date and time at which an event takes place.
TRA	See Test risk analysis
Transition	Changeover
Triage meeting	See Error management
UML	Unified Modeling Language This is a model-based language that was designed so that object-oriented analyses and designs can be made for information systems.
Use case	Specification technique that originated in UML. User scenarios are described as a sequence of transactions. Describing the dialogue between a user and the system with a tangible result. After [ISTQB, 2007].
User	Person who uses one or more applications to perform his daily activities. There is a difference between the customer or end-user and the users in the organization.

User acceptance test
(UAT)

Validation test (fit for purpose) that checks whether the users can use the system and whether the system is well integrated with the workflow and processes.

Walk through

A step-by-step presentation by the author of a document in order to gather information and to establish a common understanding of its content [ISTQB, 2007] In this book primairy used as a tour through the test design whereby the test coordinator describes how the team approached the test design and how the risks were included.

White-box test

Testing based on an analysis of the internal structure of the component or system [ISTQB, 2007].

Work breakdown structure
(WBS)

Planning technique that beaks the project down into little pieces. Each piece consists of a product or an activity. An estimate is provided of the number of hours needed for each activity.

Workaround

An alternative solution that ensures that the original goal is achieved. In software development, the term is used to specify that the function is not working properly due to a software error, but that there is a way to work around the problem. This means that the impact of the error is small.

References

[Akamai, 2006] Akamai technologies and JupiterResearch, *4 Seconds as the New Threshold of Acceptability for Retail Web Page Response Times,* Cambridge, (2006)

[Anderson, 1999] Anderson, M, *The top 13 mistakes in load testing applications,* STQmagazine, Vol. 1, No. 5, SQE Pub, (1999)

[Ash, 2006] Edwin van Ash, *Succes van SOA staat of valt met beheer,* IT Beheer, issue 9, Sdu publishers, (2006)

[ASL] ASL, *ASL begrippenlijst (glossary),* www.aslfoundation.org/

[Baars et al., 2006] Cor Baars, Gert Florijn, *Practische Enterprise Architectuur,* Seminar "Enterprise Architectuur", (2006)

[Basili] Victor R. Basili, Scott Green, Oliver Laitenberger, Filippo Lanubile, Forrest Shull1, Sivert Sørumgård, Marvin V. Zelkowitz, *The Empirical Investigation of Perceptive-Based Reading,* whitepaper, Nasa

[BCS SIGIST, 1997] British Computer Society Specialist Interest Group in Software Testing, *BS 7925 – 2 Standard for Software Component Testing Working Draft 3.3,* (1997)

[Bilt] http://mot.vuse.vanderbilt.edu/mt322/Ishikawa.htm

[Black, 2002] Rex Black, Eurostar, (2002)

[Boehm, 1981] Boehm, B., *Software Engineering Economics,* Prentice-Hall Inc, (1981)

[Bouman, 2004] Egbert Bouman, *SmarTest,* Ten Hagen & Stam, (2004)

[Braspenning, 2006] Niels Braspenning, Asia van de Mortel-Fronczak, Koos Rooda, *Model-based integration and testing of high-tech multi-disciplinary systems,* Dutch testing day, (2006)

[Broekman et al., 2001] B. Broekman, C. Hoos, M. Paap, *Automatisering van de testuitvoering,* Ten Hagen & Stam, (2001)

[Buchholtz, 2006] Eckhard Buchholtz, *Controlling Mindsets As Key To Success In Testing,* Eurostar, (2006)

[BusinessFuture] www.businessfuture.co.uk/html/ProjectManagement/Work_Breakdown_Structure.html

[Clermont, 2006] Markus Clermont, *Surviving in a QA-Organisation,* Dutch testing day, (2006)

[Computerwoorden] www.computerwoorden.nl

[Dale] Van Dale, *One of leading Dutch dictionaries,*

[DeMarco, 1999] Tom DeMarco, Liste Timothy, *peopleware- productive projects and teams,* Dorset house, New York, (1999)

[Deming] www.12manage.com/methods_demingcycle_nl.html

[Dietz] Jan Dietz, *De skills van de ICT-architect,* Cibit advisers, Technical university Delft, (1996)

[Dustin et al., 2005] E. Dustin, J. Rashka, J.Paul, *Automated software testing,* Addison Wesley, (2005)

[Fewster et al., 1999] Mark Fewster, Dorothy Graham, *Software Test Automation,* Addison Wesley, (1999)

[Fewster, 2006] Mark Fewster, *Keyword-Driven Test Automation,* Eurostar, (2006)

[Gardiner, 2006] Julie Gardiner, *Risk based teststrategy,* Event by Dutch test association, (2006)

[Gerrard] Paul Gerrard, *risk based test reporting,*

[Hansche et al.] Susan Hansche, John Berti, Chris Hare, *Official (ISC)² Guide to the CISSP EXAM,* Auerbach, (2003)

[Hedeman, 2000] Drs. Ir. B.H. Hedeman, *Prince-heerlijk,* Ten Hagen & Stam, (2000)

[Herzog] Pete Herzog, *Open Source Security Testing Methodology Manual,* www.osstmm.org

[Hul, 2006] Erwin van den Hul, *De stappen van een complexe risico analyse matrix naar concreet testen, versie 1.0- 10-08-2006,* whitepaper, (2006)

[Hul2, 2006] Erwin van den Hul, *Van een complexe risicomatrix naar concreet testen,* Event by Dutch test association, (2006)

[IEEE 829, 1998] Software Engineering Technical Committee of the IEEE Computer Society, *IEEE Std 829-1998 IEEE Standard for Documentation,*

[ISEB practitioner, 2004] Improve QS, *ISEB practitioner reader part 2- spring 2004,* Reader, (2004)

[ISTQB, 2007] International Software Testing Qualifications Board, *Standard glossary of terms used in Software Testing,* Version 2.0 (dd. December, 2nd 2007), (2007)

[ITIL, 2006] ITIL, *ITIL® Glossary v01, 1 May 2006,* www.itil.co.uk/glossary.htm] (2006)

[ITSMF, 2000] *IT Service Management, een introductie v2.3,* (2000)

[Jones, 2000] Capers Jones, *Software assessments, benchmarks and best practices,* (2000)

[Kent, 2005] John Kent, *The econoomics of testautomation,* Eurostar, (2005)

[Koomen et al., 2000] Tim koomen, Martin Pol, *Test Proces Improvement,* Ten Hagen & Stam, (2000)

[Koomen et al., 2007] Koomen et al.l, *T-map Next,* Tutein Nolthenius, (2006)

[Kramers, 1987] *Kramers woordenboek,* Elsevier/Meulenhoff Educatied, (1987)

[Marselis et al., 2007] Rik Marselis, Jos van Rooyen, Chris Schotanus i.c.w.Iris Pinkster, *Test-Grip,* LogicaCMG

[Mash, 2006] Richard Mash, *Keeping your Metrics Message Simple,* Eurostar, (2006)

[Mors, 1993] N.P.M. Mors, *Beslissingstabellen,* Lansa Publisching BV, (1993)

[Ommeren, 2006] Erik van Ommeren, *De wereld op zijn kop,* IT Beheer, issue 9, Sdu publishers, (2006)

[OWASP] *OWASP Guide 3.0,* www.owasp.org

[Pareto] www.economische-begrippen.nl/pqrs.htm

[Pas, 2004] Jens Pas, *Testwijzer,* I2B, (2004)

[Pavankumar] P. Pavankumar, *Test case design methodologies (black-box methods),*

[Pinkster et al., 2004] Iris Pinkster, bob van de Burgt, Dennis Janssen, Erik van Veenendaal, *Succesful Testmanagement,* Springer, (2004)

[Pol et al., 1999] Martin Pol, Erik van Veendendaal, Ruud Teunissen, *Testen volgens T-map,* Tutein Nolthenius, (1999)

[Prince-2, 2005] Tso, *Managing Successful Projects With Prince 2,* THE STATIONERY OFFICE, (2005)

[Quentin, 2006] Geof Quentin, *Breakfast talkshow,* Eurostar, (2006)

[Sambaer, 2006] Sven Sambaer, Alec Puype, Steven Mertens, *STBox,* Computer Task group, (2006)

[Schaefer, 2004] Hans Schaefer, *What we knew about testing 10 years ago- And still don't do,* Eurostar, (2005)

[Siteur, 2000] Maurice Siteur, *Testen met testtools,* Acidemic service, (2000)

[Siteur, 2005] Maurice Siteur, *Automate your testing!,*

[Spillner et al., 2003] Andreas Spillner, Tilo Linz, Martin Pol, *Testen volgens ISEB,* Tutein Nolthenius, (2003)
[Steenberg, 2005] Bart Steenbergen, *Bluf your way into Prince Bart Steenbergen,* whitepaper, (2005)
[SurfNet] www.surfkit.nl/tools/woordenijst/c.html
[TestFrame, 1996] CMG, *Testframe,* Ten Hagen & Stam, (1999)
[Thillard, 2006] Edwin van den Thillard, *De complexiteit van SOA,* IT Beheer, issue 9 (Dutch magazin on maintenance), Sdu publishers, (2006)
[Thompson, 2004] Neil Thompson, *Risk Mitigation Trees,* Eurostar, (2004)
[Tijman, 2007] Anko Tijman, *Agile Testen- Testen als teamsport,* (2007)
[van Es et al., 2005] Van Es, Gerwen, Graave, Lighthart, Rooij, *Service-Oriented Architecture,* Sdu publishers, (2005)
[Veenendaal, 2002] Erik van Veenendaal, *The testing practitioner,* Tutein Nolthenius, (2002)
[Veenendaal, 2004] Erik van Veenendaal, *Exploratory Testen, zinvol of onzin?,* Software Release Magazine, Jaargang 9, November 2004, Nummer 7., (2004)
[Veenendaal] Erik van Veenendaal, *Exploratory testing, wat is het nu echt ?,* presentation
[Webopedia] www.webopedia.com/TERM/D/drill_down.html
[Wikipedia] http://nl.wikipedia.org
[Zambelich] Keith Zambelich, *Totally Data-Driven Automated Testing,* whitepaper
[ZBC] www.zbc.nu
[Zeist et al., 1996] Bob van Zeist Paul Hendriks Robbert Paulussen Jos Trienekens, *Quint: Kwaliteit van softwareprodukten Praktijkervaringen met een kwaliteitsmodel,* PDF-version 1.0, mei 1996, (1996)

Index